Fiberglass Pipe Design

AWWA MANUAL M45

Second Edition

American Water Works Association

Science and Technology

AWWA unites the drinking water community by developing and distributing authoritative scientific and technological knowledge. Through its members, AWWA develops industry standards for products and processes that advance public health and safety. AWWA also provides quality improvement programs for water and wastewater utilities.

MANUAL OF WATER SUPPLY PRACTICES—M45, Second Edition
Fiberglass Pipe Design

Copyright © 2005 American Water Works Association

All rights reserved. No part of this publication may be reproduced or transmitted in any form or by any means, electronic or mechanical, including photocopy, recording, or any information or retrieval system, except in the form of brief excerpts or quotations for review purposes, without the written permission of the publisher.

Disclaimer
The authors, contributors, editors, and publisher do not assume responsibility for the validity of the content or any consequences of their use. In no event will AWWA be liable for direct, indirect, special, incidental, or consequential damages arising out of the use of information presented in this book. In particular, AWWA will not be responsible for any costs, including, but not limited to, those incurred as a result of lost revenue. In no event shall AWWA's liability exceed the amount paid for the purchase of this book.

Senior Acquisitions Manager: Colin Murcray
Project Manager/Copy Editor: Mary Kay Kozyra
Produced by Glacier Publishing Services, Inc.
Cover photo courtesy of Hobas Pipe USA, Houston, Texas

Library of Congress Cataloging-in-Publication Data

Fiberglass pipe design.--2nd ed.
   p. cm. -- (AWWA manual ; M45.)
Rev. ed of: Fiberglass pipe design manual. c1996.
Includes bibliographical references and index.
ISBN 1-58321-358-9
   1. Water-pipes--Design and construction. 2. Reinforced plastics. 3. Glass fibers. I. American Water Works Association. II. Fiberglass pipe design manual. III. Series.

TA448.F53 2005
628.1'5--dc22

                                                2004062673

Printed in the United States of America.

American Water Works Association
6666 West Quincy Avenue
Denver, CO 80235-3098

Printed on recycled paper

Contents

List of Figures, vii

List of Tables, xi

Foreword, xiii

Preface, xv

Acknowledgments, xvii

Chapter 1 History and Use 1

    1.1 Introduction, 1
    1.2 History, 1
    1.3 Applications, 2
    1.4 Standards, Specifications, and Reference Documents, 2
    1.5 Terminology, 7

Chapter 2 Materials, Properties, and Characteristics 9

    2.1 General, 9
    2.2 Characteristics, 9
    2.3 The Material System, 10
    2.4 Other Components, 12
    2.5 Physical Properties, 12
    2.6 Mechanical Properties, 15

Chapter 3 Manufacturing 19

    3.1 Introduction, 19
    3.2 Filament Winding, 19
    3.3 Centrifugal Casting, 22
    Reference, 24

Chapter 4 Hydraulics 25

    4.1 Hydraulic Characteristics, 25
    4.2 Preliminary Pipe Sizing, 25
    4.3 Typical Pipe Diameters, 26
    4.4 Pressure Reduction Calculations, 27
    4.5 Head Loss in Fittings, 30
    4.6 Energy Consumption Calculation Procedure, 32
    4.7 Pressure Surge, 34
    4.8 Design Examples, 35
    References, 41

Chapter 5 Buried Pipe Design 43

    5.1 Introduction, 43
    5.2 Terminology, 43
    5.3 Design Conditions, 46
    5.4 Pipe Properties, 46
    5.5 Installation Parameters, 47
    5.6 Design Procedure, 47
    5.7 Design Calculations and Requirements, 47

5.8 Axial Loads, 67
5.9 Special Design Considerations, 67
5.10 Design Example, 67
References, 74

Chapter 6 Guidelines for Underground Installation of Fiberglass Pipe 75

6.1 Introduction, 75
6.2 Related Documents, 76
6.3 Terminology, 77
6.4 In Situ Soils, 79
6.5 Embedment Materials, 79
6.6 Trench Excavation, 83
6.7 Pipe Installation, 85
6.8 Field Monitoring, 91
6.9 Contract Document Recommendations, 92
Reference, 92

Chapter 7 Buried Pipe Thrust Restraints 93

7.1 Unbalanced Thrust Forces, 93
7.2 Thrust Resistance, 94
7.3 Thrust Blocks, 95
7.4 Joints With Small Deflections, 97
7.5 Restrained (Tied) Joints, 99

Chapter 8 Aboveground Pipe Design and Installation 105

8.1 Introduction, 105
8.2 Thermal Expansion and Contraction, 105
8.3 Thermal Expansion Design, 106
8.4 Supports, Anchors, and Guides, 111
8.5 Bending, 116
8.6 Thermal Conductivity, 117
8.7 Heat Tracing, 117
8.8 Characteristics and Properties, 118
8.9 Design Examples, 120

Chapter 9 Joining Systems, Fittings, and Specials 125

9.1 Introduction, 125
9.2 Fiberglass Pipe Joining Systems Classification, 125
9.3 Gasket Requirements, 126
9.4 Joining Systems Description, 126
9.5 Assembly of Bonded, Threaded, and Flanged Joints, 132
9.6 Fittings and Specials, 135
9.7 Service Line Connections, 137
Reference, 137

Chapter 10 Shipping, Handling, Storage, and Repair 139

    10.1 Introduction, 139
    10.2 Shipping, 139
    10.3 Handling, 140
    10.4 Storage, 142
    10.5 Repair, 143

Glossary, 147

Index, 153

List of AWWA Manuals, 159

This page intentionally blank.

Figures

2-1 Typical circumferential stress–strain curves, 15

2-2 Typical axial stress–strain curves, 16

2-3 Static vs. cyclic pressure testing, 16

3-1 Filament winding process, 20

3-2 Application of impregnated glass reinforcement of a filament-wound pipe, 20

3-3 Continuous advancing mandrel method, 21

3-4 Finished pipe emerging from curing oven, 22

3-5 Preformed glass reinforcement sleeve method, 22

3-6 Chopped glass reinforcement method, 23

3-7 Application of glass, resin, and sand, 23

4-1 Friction pressure loss due to water flow through fiberglass pipe, 27

4-2 Moody diagram for determination of friction factor for turbulent flow, 31

5-1 Distribution of AASHTO HS-20 or HS-25 live load through granular fill for $h \leq 45$ in. (1.14 m), 54

5-2 AASHTO HS-20 live load, soil load (120 pcf), and total load graph, 58

5-3 AASHTO HS-25 live load, soil load (120 pcf), and total load graph, 58

5-4 Cooper E80 live load, soil load (120 pcf), and total load graph, 59

6-1 Trench cross-section terminology, 78

6-2 Examples of bedding support, 86

6-3 Accommodating differential settlement, 87

6-4 Cross-over of adjacent piping systems, 87

6-5 Proper compaction under haunches, 89

7-1 Thrust force definitions, 94

7-2 Typical thrust blocking of a horizontal bend, 95

7-3 Typical profile of vertical bend thrust blocking, 97

7-4 Restraint of thrust at deflected joints on long-radius horizontal curves, 98

7-5 Computation diagram for earth loads on trench conduits, 100

7-6 Restraint of uplift thrust at deflected joints on long-radius vertical curves, 101

7-7 Thrust restraint with tied joints at bends, 101

7-8 Length of tied pipe on each leg of vertical (uplift) bend, 103

8-1 Typical expansion joint installation, 108

8-2 Expansion loop dimensions, 109

8-3 Directional change, 111

8-4 Guide support, 111

8-5	Anchor support, 112	
8-6	Typical support, 113	
8-7	Fiberglass wear protection cradle, 115	
8-8	Steel wear protection cradle, 116	
8-9	Vertical support, 116	
9-1	Tapered bell-and-spigot joint, 127	
9-2	Straight bell and straight spigot joint, 127	
9-3	Tapered bell and straight spigot joint, 127	
9-4	Overlay joint construction, 128	
9-5	Overlay joint, 128	
9-6	Tapered ends overlay joint, 128	
9-7	Bell-and-spigot overlay joint, 129	
9-8	Single-gasket bell-and-spigot joint, 129	
9-9	Single-gasket spigot, 129	
9-10	Double-gasket bell-and-spigot joint, 130	
9-11	Double-gasket spigot, 130	
9-12	Gasketed coupling joint, 130	
9-13	Gasketed coupling joint—cross section, 131	
9-14	Restrained-gasketed bell-and-spigot joint, 131	
9-15	Restrained-gasketed coupling joint, 131	
9-16	Restrained-gasketed threaded bell-and-spigot O-ring joint, 131	
9-17	Fiberglass flange to fiberglass and steel flange joint, 132	
9-18	Fiberglass flanges to flanged steel valve connection, 132	
9-19	Fiberglass flange with grooved face for O-ring seal, 133	
9-20	Mechanical coupling joint, 133	
9-21	Compression molded fittings, 136	
9-22	Flanged compression molded fittings, 136	
9-23	Mitered fitting configurations, 136	
9-24	Mitered fitting, 137	
9-25	Mitered fitting fabrication, 137	
9-26	Mitered fittings, 138	
9-27	Mitered fitting field fabrication, 138	
9-28	Fittings field assembly, 138	
10-1	Pipe shipment by truck, 140	
10-2	Single sling handling, 141	
10-3	Double sling handling, 141	

10-4	Unitized small-diameter bundle, 141	
10-5	Unitized load handling, 142	
10-6	Handling nested pipes, 142	
10-7	Denesting pipes, 143	
10-8	Pipe stacking, 143	
10-9	Patch, 144	
10-10	Cut out and replace, 144	
10-11	Steel coupling, 144	

This page intentionally blank.

Tables

2-1 Mechanical properties range, 14

4-1 Typical K factors for fiberglass fittings, 32

5-1 Shape factors, 51

5-2 AASHTO HS-20, HS-25, and Cooper E80 live loads, 57

5-3 Soil classification chart, 61

5-4 M_{sb} based on soil type and compaction condition, 62

5-5 Values for the soil support combining factor S_c, 64

5-6 Values for the constrained modulus of the native soil at pipe zone elevation, 64

5-7 Conditions and parameters for design example, 68

6-1 Soil stiffness categories, 80

6-2 Recommendations for installation and use of soils and aggregates for foundation and pipe zone embedment, 81

6-3 Maximum particle size for pipe embedment, 82

7-1 Horizontal soil-bearing strengths, 96

8-1 Minimum support width for 120° contact supports, 115

This page intentionally blank.

Foreword

The American Water Works Association prepares documents, including manuals, for water supply service applications. Chapters 1 and 2 of this manual contain general information about applications other than water supply service for fiberglass pipe for informational and historical purposes. The use of this manual is intended for water supply service applications.

This page intentionally blank.

Preface

This is the second edition of AWWA Manual M45, *Fiberglass Pipe Design*. This manual provides the reader with both technical and general information to aid in the design, specification, procurement, installation, and understanding of fiberglass pipe and fittings. It is a discussion of recommended practice, not an AWWA standard calling for compliance with certain specifications. It is intended for use by utilities and municipalities of all sizes, whether as a reference book or textbook for those not fully familiar with fiberglass pipe and fitting products. Design engineers and consultants may use this manual in preparing plans and specifications for new fiberglass pipe design projects.

The manual covers fiberglass pipe and fitting products and certain appurtenances and their application to practical installations, whether of a standard or special nature. For adequate knowledge of these products, the entire manual should be studied. Readers will also find the manual a useful source of information when assistance is needed with specific or unusual conditions. The manual contains a list of applicable national standards, which may be purchased from the respective standards organizations (e.g., American Water Works Association, American Society for Testing and Materials, etc.).

This page intentionally blank.

Acknowledgments

The American Water Works Association (AWWA) Fiberglass Pipe Design Manual Subcommittee, which developed this manual, had the following personnel at the time:

Richard C. Turkopp, *Chair*

A.B. Colthorp, Lake St. Louis, Mo.	(AWWA)
R.P. Fuerst, U.S. Bureau of Reclamation, Denver, Colo.	(AWWA)
N.E. Kampbell, Inliner Technologies, Paoli, Ind.	(AWWA)
David Kozman, Rinker-Pipeline Renewal, Hilliard, Ohio	(AWWA)
Jim Loeffler, Interplastics Corp., Minneapolis, Minn.	(AWWA)
A.M. May, Little Rock, Ark.	(AWWA)
William McCann, Amitech America, Zachary, La.	(AWWA)
T.J. McGrath, Simpson Gumpertz & Heger Inc., Waltham, Mass.	(AWWA)
Lee Pearson, Vero Beach, Fla.	(AWWA)
P.A. Sharff, Simpson Gumpertz & Heger Inc., Waltham, Mass.	(AWWA)
Rich Stadelman, Reichhold Chemicals, Durham, N.C.	(AWWA)
Rick Turkopp, Hobas Pipe USA, Houston, Texas	(AWWA)

This manual was also reviewed and approved by the AWWA Standards Council and the Standards Committee on Thermosetting Fiberglass Reinforced Plastic Pipe. The Standards Committee on Thermosetting Fiberglass Reinforced Plastic Pipe had the following personnel at the time of approval:

Timothy J. McGrath, *Chair*

Consumer Members

P.A. Fragassi, Lake County Public Water District, Zion, Ill.	(AWWA)
R.P. Fuerst, U.S. Bureau of Reclamation, Denver, Colo.	(AWWA)
K.W. Kells, Ivoryton, Conn.	(AWWA)

General Interest Members

S.J. Abrera Jr., South Pasadena, Calif.	(AWWA)
J.H. Bambei Jr.,[*] Denver Water, Denver, Colo.	(AWWA)
J.P. Biro, Houston, Texas	(AWWA)
J.K. Jeyapalan, Pipe Consultant, New Milford, Conn.	(AWWA)
R.A. Johnson, Russcor Engineering, Naples, Fla.	(AWWA)
N.E. Kampbell, Inliner Technologies, Paoli, Ind.	(AWWA)
T.J. McGrath, Simpson Gumpertz & Heger Inc., Waltham, Mass.	(AWWA)
P.J. Olson,[*] Staff Engineer Liaison, AWWA, Denver, Colo.	(AWWA)

---

[*]Liaison, nonvoting

Producer Members

William McCann, Amitech America Ltd., Zachary, La.	(AWWA)
R.I. Mueller, Ameron International, Rancho Cucamonga, Calif.	(AWWA)
Lee Pearson, Vero Beach, Fla.	(AWWA)
Rick Turkopp, Hobas Pipe USA, Houston, Texas	(AWWA)

AWWA MANUAL M45

Chapter 1

History and Use

1.1 INTRODUCTION

Fiberglass pipe is made from glass fiber reinforcements embedded in, or surrounded by, cured thermosetting resin. This composite structure may also contain aggregate, granular, or platelet fillers; thixotropic agents; and pigments or dyes. By selecting the proper combination of resin, glass fibers, fillers, and design, the fabricator can create a product that offers a broad range of properties and performance characteristics. Over the years, the diversity and versatility of materials used to manufacture fiberglass pipe have led to a variety of names for fiberglass pipe. Among these are reinforced thermosetting resin pipe (RTRP), reinforced polymer mortar pipe (RPMP), fiberglass reinforced epoxy (FRE), glass reinforced plastic (GRP), and fiberglass reinforced plastic (FRP). Fiberglass pipes have also been categorized by the particular manufacturing process—filament winding or centrifugal casting. Frequently, the particular resin used to manufacture the fiberglass pipe—epoxy, polyester, or vinyl ester—has been used to classify or grade fiberglass pipes.

Regardless of the many possible combinations, the most common and useful designation is simply "fiberglass pipe." This name encompasses all of the various available products and allows consideration as a unique and general class of engineering materials.

1.2 HISTORY

Fiberglass pipe was introduced in 1948. The earliest application for fiberglass piping, and still one of the most widely used, is in the oil industry. Fiberglass pipe was selected as a corrosion-resistant alternative to protected steel, stainless steel, and other more exotic metals. Product lines expanded to include applications of increasingly high pressure and down-hole tubing with threaded connections. In the late 1950s, larger diameters became available and fiberglass pipe was increasingly used in the chemical process industry because of the pipe's inherent corrosion-resistant characteristics.

Since the 1960s, fiberglass pipe products have been used for municipal water and sewage applications. Fiberglass pipe combines the benefits of durability, strength, and corrosion resistance, thus eliminating the need for interior linings, exterior coatings, and cathodic protection. Fiberglass pipe systems offer great design flexibility with a wide range of standard pipe diameters and fittings available, as well as an inherent ability for custom fabrication to meet special needs. Fiberglass pipe is available in diameters ranging from 1 in. through 144 in. (25 mm through 3,600 mm). Fiberglass pipe is available in pressure classes ranging from gravity applications through several thousand per square inch (kilopascals). There are few countries in the world where fiberglass pipe has not been used.

1.3 APPLICATIONS

Fiberglass pipe is used in many industries and for a myriad of applications, including:

- chemical processes,
- desalination,
- down-hole tubing and casing,
- ducting and vent piping,
- geothermal,
- industrial effluents,
- irrigation,
- oil fields,
- potable water,
- power plant cooling and raw water,
- sanitary sewers,
- seawater intake and outfalls,
- slurry piping,
- storm sewers,
- water distribution, and
- water transmission.

1.4 STANDARDS, SPECIFICATIONS, AND REFERENCE DOCUMENTS

Many organizations have published nationally recognized standards, test methods, specifications, and recommended practices on fiberglass piping systems and products. These organizations include the American Society for Testing and Materials (ASTM), the American Petroleum Institute (API), the American Society of Mechanical Engineers (ASME), the NSF International (NSF), Underwriters Laboratories (UL), Factory Mutual Research (FM), the American National Standards Institute (ANSI), and the International Organization for Standardization (ISO).

Following is a listing of fiberglass pipe standards and specifications that are commonly used in specifying, testing, and using fiberglass piping systems.

1.4.1 Product Specifications and Classifications

General

ASTM D2310	Standard Classification for Machine-Made "Fiberglass" (Glass-Fiber-Reinforced Thermosetting-Resin) Pipe
ASTM D2517	Standard Specification for Reinforced Epoxy Resin Gas Pressure Pipe and Fittings
ASTM D2996	Standard Specification for Filament-Wound "Fiberglass" (Glass-Fiber-Reinforced Thermosetting-Resin) Pipe (Applicable to epoxy, polyester, and furan resins in sizes from 1 in. to 16 in. [25 mm to 400 mm].)
ASTM D2997	Standard Specification for Centrifugally Cast "Fiberglass" (Glass-Fiber-Reinforced Thermosetting-Resin) Pipe (Applicable for 1 in. through 14 in. [25 mm through 350 mm] pipe of polyester or epoxy resins.)
ASTM D3262	Standard Specification for "Fiberglass" (Glass-Fiber-Reinforced Thermosetting-Resin) Sewer Pipe (Applicable for pipes 8 in. through 144 in. [200 mm through 3,700 mm] diameter, with or without siliceous sand, and polyester or epoxy resin.)
ASTM D3517	Standard Specification for "Fiberglass" (Glass-Fiber-Reinforced Thermosetting-Resin) Pressure Pipe (Applicable for pipes 8 in. through 144 in. [200 mm through 3,700 mm] diameter, with or without siliceous sand, and polyester or epoxy resin.)
ASTM D3754	Standard Specification for "Fiberglass" (Glass-Fiber-Reinforced Thermosetting-Resin) Sewer and Industrial Pressure Pipe (Applicable for 8 in. through 144 in. [200 mm through 3,700 mm] diameter, with or without siliceous sand, and polyester or epoxy resin.)
ASTM D4024	Standard Specification for Machine Made "Fiberglass" (Glass-Fiber-Reinforced Thermosetting-Resin) Flanges (Applicable for ½ in. through 24 in. [13 mm through 600 mm] ANSI B16.5 150 lb [70 kg] bolt circle flanges.)
ASTM D4161	Standard Specification for "Fiberglass" (Glass-Fiber-Reinforced Thermosetting-Resin) Pipe Joints Using Flexible Elastomeric Seals
ASTM F1173	Standard Specification for Thermosetting Resin Fiberglass Pipe Systems to Be Used for Marine Applications
API 15LR	Specification for Low Pressure Fiberglass Line Pipe (Applicable to 2 in. through 24 in. [50 mm through 600 mm] diameter pipe of epoxy or polyester resin for use at cyclic pressures to 1,000 psi [6,895 kPa].)
API 15HR	Specification for High Pressure Fiberglass Line Pipe (Applicable to 1 in. through 10 in. [25 mm through 250 mm] pipe and fittings for operating pressures of 500 psi [3,500 kPa] to 5,000 psi [35,000 kPa].)
ANSI/AWWA C950	AWWA Standard for Fiberglass Pressure Pipe

US military specifications

MIL P24608 — Specification for epoxy resin pipe from $\frac{1}{2}$ in. through 12 in. (13 mm through 300 mm) diameters for 200 psig (1,379 kPa) service at 150°F (66°C) for US Navy shipboard applications

MIL P28584A — Specification for epoxy resin pipe and fittings from 2 in. through 12 in. (50 mm through 300 mm) diameter for use as Steam Condensate Return Lines in continuous service at 125 psig (862 kPa) and 250°F (121°C)

MIL P29206A — Specification for epoxy or polyester pipe and fittings 2 in. through 12 in. (50 mm through 300 mm) in diameter for POL services to 150°F (66°C) and 150 psig (1,034 kPa) with surges to 250 psig (1,724 kPa)

1.4.2 Recommended Practices

Dimensions

ASTM D3567 — Standard Practice for Determining Dimensions of "Fiberglass" (Glass-Fiber-Reinforced Thermosetting-Resin) Pipe and Fittings

Installation

ASTM D3839 — Standard Guide for Underground Installation of "Fiberglass" (Glass-Fiber-Reinforced Thermosetting-Resin) Pipe

API RP15TL4 — Care and Use of Fiberglass Tubulars

API RP1615 — Installation of Underground Petroleum Storage Systems

1.4.3 Standard Test Methods

Tensile properties

ASTM D638 — Standard Test Method for Tensile Properties of Plastics

ASTM D1599 — Standard Test Method for Resistance to Short-Time Hydraulic Failure Pressure of Plastic Pipe, Tubing and Fittings

ASTM D2105 — Standard Test Method for Longitudinal Tensile Properties of "Fiberglass" (Glass-Fiber-Reinforced Thermosetting-Resin) Pipe and Tube

ASTM D2290 — Standard Test Method for Apparent Hoop Tensile Strength of Plastic or Reinforced Plastic Pipe by Split Disk Method

Compressive properties

ASTM D695 — Standard Test Method for Compressive Properties of Rigid Plastics

Bending properties

ASTM D790 — Standard Test Methods for Flexural Properties of Unreinforced and Reinforced Plastics and Electrical Insulating Materials

ASTM D2925 — Standard Test Method for Beam Deflection of "Fiberglass" (Glass-Fiber-Reinforced Thermosetting-Resin) Pipe Under Full Bore Flow

Long-term internal pressure strength

ASTM D1598 — Standard Test Method for Time-to-Failure of Plastic Pipe Under Constant Internal Pressure

ASTM D2143   Standard Test Method for Cyclic Pressure Strength of Reinforced, Thermosetting Plastic Pipe

ASTM D2992   Standard Practice for Obtaining Hydrostatic or Pressure Design Basis for "Fiberglass" (Glass-Fiber-Reinforced Thermosetting-Resin) Pipe and Fittings

Pipe stiffness

ASTM D2412   Standard Test Method for Determination of External Loading Characteristics of Plastic Pipe by Parallel-Plate Loading

External pressure

ASTM D2924   Standard Test Method for External Pressure Resistance of "Fiberglass" (Glass-Fiber-Reinforced Thermosetting-Resin) Pipe

Chemical resistance

ASTM C581   Standard Practice for Determining Chemical Resistance of Thermosetting Resins Used in Glass-Fiber-Reinforced Structures Intended for Liquid Service

ASTM D3681   Standard Test Method for Chemical Resistance of "Fiberglass" (Glass-Fiber-Reinforced Thermosetting-Resin) Pipe in a Deflected Condition

ASTM D5365   Standard Test Method for Long-Term Ring-Bending Strain of "Fiberglass" (Glass-Fiber-Reinforced Thermosetting-Resin) Pipe

1.4.4 Product Listings, Approvals, and Piping Codes

NSF International—Standard Numbers 14 and 61. Tests and lists fiberglass pipe, fittings, and adhesives for use in conveying potable water. Additionally tests and certifies products as to their classification to an applicable national standard or for special properties (Standard 14 only).

Underwriters Laboratories, Inc. Provides established standards for testing and listing fiberglass pipe for use as underground fire water mains and underground transport of petroleum products.

Factory Mutual Research. Has established an approval standard for plastic pipe and fittings for underground fire protection service.

ANSI/ASME B31.1—Power Piping Code. This code prescribes minimum requirements for the design, materials, fabrication, erection, testing, and inspection of power and auxiliary service piping systems for electric generation stations, industrial institutional plants, and central and district heating plants.

ANSI/ASME B31.3—Chemical Plant and Petroleum Refinery Piping Code. This code lists some ASTM, AWWA, and API fiberglass pipe specifications as acceptable for use within the code and establish criteria for their installation and use. These codes, in addition to other ASME codes, establish rules regarding the application of fiberglass piping and provide engineering guidance for the use of fiberglass materials.

ANSI/ASME B31.8—Gas Transmission and Distribution Piping Systems Code. This code lists fiberglass pipe manufactured in compliance with ASTM D2517 as acceptable for use within the code.

Department of Transportation, Title 49, Part 192. This is a code of federal regulations that covers the transportation of natural and other gases by pipeline. Minimum federal standards are included.

ASME Boiler and Pressure Vessel Code Case N155. This code provides the rules for the construction of fiberglass piping systems for use in section III, division I, class 3 applications in nuclear power plants.

1.4.5 International Organization for Standardization Standards and Specifications

ISO has issued many standards, test methods, and technical reports relating to fiberglass piping systems and products. Many of their titles, as well as the general content, are very similar to the US-issued standards covered previously.

Product specifications

ISO 10467	Plastics piping systems for pressure and non-pressure drainage and sewerage—Glass-reinforced thermosetting plastics (GRP) systems based on unsaturated polyester (UP) resin
ISO 10639	Plastics piping systems for pressure and non-pressure water supply—Glass-reinforced thermosetting plastics (GRP) systems based on unsaturated polyester (UP) resin

Test methods

ISO 7432	Glass-reinforced thermosetting plastics (GRP) pipes and fittings—Test methods to prove the design of locked socket-and-spigot joints, including double-socket joints, with elastomeric seals
ISO 7509	Plastics piping systems—Glass-reinforced thermosetting plastics (GRP) pipes—Determination of time to failure under sustained internal pressure
ISO 7510	Plastics piping systems—Glass-reinforced thermosetting plastics (GRP) components —Determination of the amounts of constituents using the gravimetric method
ISO 7511	Plastics piping systems—Glass-reinforced thermosetting plastics (GRP) pipes and fittings—Test methods to prove the leaktightness of the wall under short-term internal pressure
ISO 7684	Plastics piping systems—Glass-reinforced thermosetting plastics (GRP) pipes—Determination of the creep factor under dry conditions
ISO 7685	Plastics piping systems—Glass-reinforced thermosetting plastics (GRP) pipes—Determination of initial specific ring stiffness
ISO 8483	Glass-reinforced thermosetting plastics (GRP) pipes and fittings—Test methods to prove the design of bolted flanged joints
ISO 8513	Plastics piping systems—Glass-reinforced thermosetting plastics (GRP) pipes—Determination of longitudinal tensile properties
ISO 8521	Plastics piping systems—Glass-reinforced thermosetting plastics (GRP) pipes—Determination of the apparent initial circumferential tensile strength
ISO 8533	Glass-reinforced thermosetting plastics (GRP) pipes and fittings—Test methods to prove the design of cemented or wrapped joints
ISO 8639	Glass-reinforced thermosetting plastics (GRP) pipes and fittings—Test methods for leaktightness of flexible joints

ISO 10466	Plastics piping systems—Glass-reinforced thermosetting plastics (GRP) pipes—Test method to prove the resistance to initial ring deflection
ISO 10468	Glass-reinforced thermosetting plastics (GRP) pipes—Determination of the long-term specific ring creep stiffness under wet conditions and the calculation of the wet creep factor
ISO 10471	Glass-reinforced thermosetting plastics (GRP) pipes—Determination of the long-term ultimate bending strain and the long-term ultimate relative ring deflection under wet conditions
ISO 10928	Plastics piping systems—Glass-reinforced thermosetting plastics (GRP) pipes and fittings—Methods for regression analysis and their use
ISO 10952	Plastics piping systems—Glass-reinforced thermosetting plastics (GRP) pipes and fittings—Determination of the resistance to chemical attack from the inside of a section in a deflected condition
ISO 14828	Glass-reinforced thermosetting plastics (GRP) pipes—Determination of the long-term specific ring relaxation stiffness under wet conditions and the calculation of the wet relaxation factor
ISO 15306	Glass-reinforced thermosetting plastics (GRP) pipes—Determination of the resistance to cyclic internal pressure

Technical reports

ISO/TR 10465-1	Underground installation of flexible glass-reinforced thermosetting resin (GRP) pipes—Part 1: Installation procedures
ISO/TR 10465-2	Underground installation of flexible glass-reinforced thermosetting resin (GRP) pipes—Part 2: Comparison of static calculation methods
ISO/TR 10465-3	Underground installation of flexible glass-reinforced thermosetting resin (GRP) pipes—Part 3: Installation parameters and application limits

1.5 TERMINOLOGY

Fiberglass pipe users may encounter some unique or unfamiliar terminology. A glossary of terms used in this manual and by those in the fiberglass pipe industry is provided at the end of this manual.

This page intentionally blank.

AWWA MANUAL M45

Chapter 2

Materials, Properties, and Characteristics

2.1 GENERAL

Fiberglass pipe is a composite material system produced from glass fiber reinforcements, thermosetting plastic resins, and additives. By selecting the right combination and amount of materials and the specific manufacturing process, the designer can create a product to meet the most demanding requirements. The result is a material with a broad range of characteristics and performance attributes.

2.2 CHARACTERISTICS

The following is a list of general characteristics of fiberglass composite pipe.

Corrosion resistance. Fiberglass pipe systems are resistant to corrosion, both inside and out, in a wide range of fluid-handling applications. As a result, additional linings and exterior coatings are not required.

Strength-to-weight ratio. Fiberglass composite piping systems have excellent strength-to-weight properties. The ratio of strength per unit of weight of fiberglass composites is greater than that of iron, carbon, and stainless steels.

Lightweight. Fiberglass composites are lightweight. Fiberglass piping is approximately one-sixth the weight of similar steel products and one-tenth the weight of similar concrete products.

Electrical properties. Standard fiberglass pipes are nonconductive. Some manufacturers offer conductive fiberglass piping systems for applications that require dissipation of static electricity buildup when transporting certain fluids, such as jet fuel.

Dimensional stability. Fiberglass composites can maintain the critical tolerances required of the most demanding structural and piping applications. The material meets the most stringent material stiffness, dimensional tolerance, weight, and cost criteria.

Low maintenance cost. Fiberglass piping is easy to maintain because it does not rust, is easily cleaned, and requires minimal protection from the environment.

2.3 THE MATERIAL SYSTEM

Fiberglass composites consist of glass fiber reinforcements, thermosetting resins, and additives, which are designed and processed to meet specific functional performance criteria. To aid understanding of the performance characteristics of a finished fiberglass pipe, the interrelationship of the system components is outlined in this chapter. The following is a list of terms used in describing the material system.

Fiberglass reinforcement. The amount, type, location, and orientation of glass fibers in the pipe that will provide the required mechanical strength.

Resin system. Resin selection will provide the physical and chemical properties (e.g., the glass transition temperature, a measurement of resistance to heat, and softening or plasticization by solvents and gases).

Following is a brief review of the constituents of fiberglass pipe and how they influence the finished pipe product.

2.3.1 Glass Fiber Reinforcements

The mechanical strength of fiberglass pipe depends on the amount, type, and arrangement of glass fiber reinforcement. Strength increases proportionally with the amount of glass fiber reinforcement. The quantity of the glass fibers (and the direction in which the individual strands are placed) determines the strength.

2.3.1.1 Fiberglass Types

Fiberglass materials are available with a variety of compositions. This allows for additional design flexibility to meet performance criteria. All fiberglass reinforcement begins as individual filaments of glass drawn from a furnace of molten glass. Many filaments are formed simultaneously and gathered into a "strand." A surface treatment (sizing) is added to maintain fiber integrity, establish compatibility with resin, and ease further processing by improving consolidation and wet strength. Sizing can also affect resin chemistry and laminate properties.

The glass fibers most commonly used in pipe are referred to as Types E, ECR, and C. Glass types ECR and C provide improved acid and chemical resistance. Type C glass fibers are generally only used to reinforce chemical-resistant liners.

2.3.1.2 Fiberglass Reinforcement Forms

Following is a brief description of the various forms of fiberglass reinforcements.

Continuous roving. These consist of bundled, untwisted strands of glass fiber reinforcement and come as cylindrical packages for further processing. Continuous roving typically is used in filament winding and unidirectional/bidirectional reinforcements and may be processed into chopped strand mat used to provide multidirectional reinforcement in pipe and fittings.

Woven roving. This is a heavy, drapable fabric, woven from continuous roving. It is available in various widths, thicknesses, and weights. Woven roving provides high strength to large molded parts and is lower in cost than conventional woven fabrics.

Reinforcing mats. These are chopped strands held together with resinous binders. There are two kinds of reinforcing mats used in pipe and fittings (i.e., chopped strand mat and woven roving combination mat). Chopped strand mats are used in medium-strength applications for pipe fittings and reinforcing where a uniform cross section is desired. Use of the combination mat saves time in hand lay-up operations.

Surface veils. These lightweight fiberglass reinforcement mats allow layers with a high resin content with minimal reinforcement. The surface veil provides extra environmental resistance for pipe and fittings, plus a smooth appearance. (Some surface veils from polyester fibers are also used.)

2.3.1.3 Reinforcement Arrangement

The three general types of fiber orientation include:

Unidirectional. The greatest strength is in the direction of the fibers. Up to 80% reinforcement content by weight is possible.

Bidirectional. Some fibers are positioned at an angle to the rest of the fibers, as with helical filament winding and woven fabrics. This provides different strength levels governed by the fiber quantity in each direction of fiber orientation. A combination of continuous and chopped fibers is also used to provide designed directional strength.

Multidirectional (isotropic). This arrangement provides nearly equal, although generally lower, strength and modulus in all directions. From 10% to 50% reinforcement content, by weight, can be obtained with multidirectional materials such as chopped roving or chopped strand mat.

2.3.2 Resins

The second major component of fiberglass pipe is the resin system. Manufacturers choose a resin system for chemical, mechanical, and thermal properties and processability.

The two basic groups of resin systems are thermosetting and thermoplastic. Fiberglass pipe, by definition, uses only thermosetting resin systems. Thermosets are polymeric resin systems cured by heat or chemical additives. Once cured, a thermoset is essentially infusible (cannot be remelted) and insoluble.

The thermosetting resins used in fiberglass pipe fall into two general categories—polyesters and epoxies.

2.3.2.1 Polyester Resins

Polyester resins are commonly used to produce large-diameter water and sewage piping. Polyesters have excellent water and chemical resistance and are noted for acid resistance.

The base polyester resin is a solid. It is typically dissolved in styrene monomer, with which it cross-links to provide the final thermoset structure. Polyester resins are cured by organic peroxide catalysts. The type and amount of catalyst will influence gel time, cure time, curing temperature, and the degree of cure. Typical catalysts include methyl ethyl ketone peroxide (MEKP) and benzyl peroxide (BPO).

Manufacturers may select from several different types of polyester resins that provide a wide range of performance characteristics. These include:

- orthophthalic polyester,
- terephthalic polyester,
- chlorendic acid polyester,
- novolac epoxy vinyl ester,
- isophthalic polyester,
- bisphenol-A fumarate polyester, and
- bisphenol-A vinyl ester.

2.3.2.2 Epoxy Resins

Epoxy resins are commonly used in the manufacture of smaller diameter piping (<30 in. [800 mm]) conveying water, condensates, hydrocarbons, caustics, and dilute acids.

Fiberglass epoxy piping is used in oil fields at pressures up to several thousand per square inch (kilopascals). Epoxy resins typically allow higher service temperatures than polyester resins, ranging up to about 225°F (108°C).

Epoxy resins cannot be categorized by resin type as easily as polyesters. The type of curing agent, or hardener, is critical with epoxy resins because the agent influences the composite properties and performance. The two basic types are amine- and anhydride-cured bisphenol-A epoxies.

Bisphenol-A epoxy resins are commonly cured with multifunctional primary amines. For optimum chemical resistance, these mixtures usually require a heat cure and/or post cure. The cured resin has good chemical resistance, particularly in alkaline environments, and can have good temperature resistance. Bisphenol-A epoxy resins may also be cross-linked with various anhydrides by using a tertiary amine accelerator and heat. These cured polymers generally have good chemical resistance, especially to acids.

2.4 OTHER COMPONENTS

Glass fiber reinforcements and thermosetting resins are the major constituents in fiberglass pipe. However, other materials that influence processing and/or product performance are used, including fillers, promoters, accelerators, inhibitors, and pigments.

Fillers. Inorganic materials, such as hydrated alumina, glass microspheres, clay, talc, calcium carbonate, sand, and calcium silicate, may yield economic, appearance, or performance advantages in fiberglass pipe.

Promoters, accelerators, and inhibitors. Promoters and accelerators advance the action of the catalyst to reduce the processing time. Inhibitors provide control over the cure cycle and increase the shelf life of the resin mix.

Pigments. The pigment choice affects the difference in reflected and transmitted color, clarity of the resin mix, reaction between dyes and other additives, such as catalysts, and the end-product color fastness and heat resistance.

2.5 PHYSICAL PROPERTIES

The following is a description of the physical properties of fiberglass pipe.

2.5.1 Chemical Resistance

All fiberglass pipes provide excellent resistance to water and native ground conditions. They are not subject to general corrosion attack, galvanic corrosion, aerobic corrosion, pitting, dezincification, and graphitic and intergranular corrosion. Fiberglass pipes are subject to some environmental stress and aging effects, the determination of which is part of the fiberglass pipe design procedure (see chapter 5).

Fiberglass pipe resists a wide range of chemicals. The chemical resistance of fiberglass pipe depends primarily on the particular resin matrix material used. Although other factors such as liner construction, cure, and fabrication method may influence the chemical resistance of fiberglass pipe, the primary factor is the resin. The resins can be selected to provide chemical resistance to a broad range of materials. The fiberglass pipe manufacturer should be consulted for performance information for a particular chemical application.

2.5.2 Temperature Resistance

The temperature resistance of fiberglass pipe also depends largely on the resin matrix. The allowable upper limit of service temperature will also be influenced by the chemical environment and the stress condition of the piping system. In general, chemical agents are more aggressive at higher concentrations and elevated temperatures. However, for typical temperatures encountered in water supply systems (33°F to 90°F [1°C to 32°C]), fiberglass pipe is unaffected by service temperature, and there is no need to rerate or derate fiberglass pipe pressure performance. Fiberglass pipe is virtually unaffected by colder temperatures. Therefore, normal shipping, handling, and storage procedures, as discussed in chapter 10, may be used in subzero weather. However, users and installers of fiberglass pipe should be aware that the coefficient of thermal expansion for fiberglass pipe is generally higher than that for metal pipes (see Table 2-1). This must be recognized and provisions made in design and installation to accommodate expansion and contraction, particularly in aboveground applications.

2.5.3 Abrasion Resistance

Fiberglass pipe provides generally good abrasion resistance and can be custom made for extremely abrasive service by lining the pipe with sand, silica flour, carborundum, or ceramic beads or tiles or by incorporating resilient liner materials such as polyurethanes. Special lining materials should match or exceed the hardness and abrasiveness of the contents being transported through the pipe or provide a high level of toughness and resilience.

2.5.4 Flame Retardants

The thermosetting resin systems used to fabricate fiberglass pipes are organic materials. Therefore, under the proper combination of heat and oxygen, a thermosetting resin, like any organic matter, will burn. If required, the fire performance of fiberglass pipe can be enhanced by using resin systems that contain halogens or phosphorus. Use of hydrated fillers also enhances flame resistance. Other additives, primarily antimony oxides, can also increase the effectiveness of halogenated resins.

Fire performance testing requires small samples and specialized test methods and may not indicate how a material will perform in a full-scale field or fire situation. The fiberglass pipe manufacturer should be consulted for specific information on the combustion performance of fiberglass pipe.

2.5.5 Weathering Resistance

Most thermosetting resin systems used to fabricate fiberglass pipe are subject to some degradation from ultraviolet (UV) light. This degradation, however, is almost entirely a surface phenomenon. The structural integrity of fiberglass pipe is not affected by exposure to UV light. The use of pigments, dyes, fillers, or UV stabilizers in the resin system or painting of exposed surfaces can help reduce significantly any UV surface degradation. Surfaces exposed to UV light are generally fabricated with a resin-rich layer. Other weathering effects, such as rain or saltwater, are resisted fully by the inherent corrosion resistance of fiberglass pipe.

2.5.6 Resistance to Biological Attack

Fiberglass pipe will not deteriorate or break down under attack from bacteria or other microorganisms, nor will it serve as a nutrient to microorganisms, macroorganisms, or fungi. There are no known cases in which fiberglass pipe products suffered degradation or deterioration due to biological action. No special engineering or installation procedures are required to protect fiberglass pipe from biological attack. Some elastomers used in gaskets may be susceptible to this type of attack.

Table 2-1 Mechanical properties range

Property	Units		Resin		Glass Fiber		Fiberglass Pipe	
	in.-lb	SI	in.-lb	SI	in.-lb	SI	in.-lb	SI
Tensile strength	10^3 psi	MPa	9.0–13.0	62–90	250–350	1,725–2,400	2.0–80.0	14–550
Tensile modulus	10^6 psi	GPa	0.4–0.6	2.8–4.1	10.0–11.0	69–76	0.5–5.0	3.5–34.5
Flexural strength	10^3 psi	MPa	10.0–22	69–150	250–350	1,725–2,400	4.0–70.0	28–480
Flexural modulus	10^6 psi	GPa	0.4–0.6	2.8–4.1	10.0–11.0	69–76	1.0–5.0	6.9–34.5
Coefficient of thermal expansion	in./in./°F×10^{-6}	mm/mm/°C×10^{-6}	11.0–55	20–100	3.0–3.3	5.4–6.0	8.0–30.0	14–54
Specific gravity			1.0–1.2	1.0–1.2	2.57–2.63	2.57–2.63	1.2–2.3	1.2–2.3
Compressive strength	10^3 psi	MPa	10.0–22	69–150	250–350	1,725–2,400	10.0–40.0	69–275

2.5.7 Tuberculation

Soluble encrustants, such as calcium carbonate, in some water supplies do not tend to precipitate onto the smooth walls of fiberglass pipe. Because fiberglass pipe is inherently corrosion resistant, there is no tuberculation of the fiberglass pipe caused by corrosion by-products.

2.6 MECHANICAL PROPERTIES

2.6.1 Mechanical Property Range

The design flexibility inherent with glass fiber reinforced plastic materials and the range of manufacturing processes used precludes the simple listing of fiberglass pipe mechanical properties. For this reason, fiberglass pipe product standards are based on performance and detail product performance requirements rather than thickness-property tables. Table 2-1 illustrates the broad range of mechanical properties available for resin, glass fiber, and fiberglass pipe.

This broad range of mechanical properties is further illustrated by the widely variable stress–strain curves possible with fiberglass pipe, depending on the amount, type, and orientation of the reinforcement as well as the manufacturing process. Figures 2-1 and 2-2 show the typical shape of the stress–strain curves for high- and low-pressure pipes for the circumferential and axial directions, respectively.

2.6.2 Mechanical Property Testing

The time dependence and wide range of mechanical properties of fiberglass pipe necessitate testing to develop data needed for design and analysis. Many test methods develop data over a moderate time range and then statistically extrapolate the data to establish long-term design values.

For example, the key long-term property test for fiberglass pipe is the development of a hydrostatic design basis (HDB) to establish the pipe pressure rating. This testing

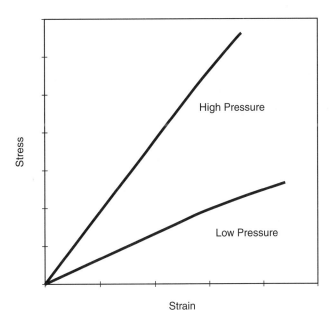

Figure 2-1 Typical circumferential stress–strain curves

16   FIBERGLASS PIPE DESIGN

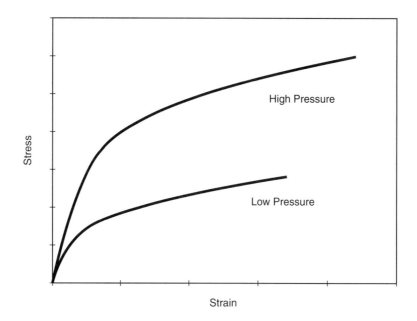

Figure 2-2 Typical axial stress–strain curves

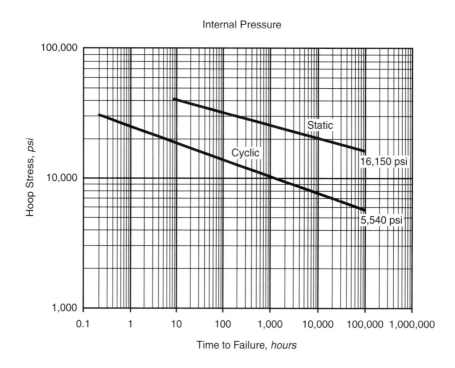

Figure 2-3 Static vs. cyclic pressure testing

is conducted in accordance with ASTM D2992. This method requires pressurizing a minimum of 18 pipe samples at pressures far exceeding the normal use range and monitoring the time to failure. Data must be collected over a range of time from 1 hour to beyond 10,000 hours. The pressure/stress/strain (all may be validly used) versus time to failure is statistically extrapolated to 50 years to establish a long-term HDB. To establish the pipe pressure rating, a safety factor is applied to this 50-year value. In ANSI/AWWA Standard C950, *Fiberglass Pressure Pipe*, the specified safety factor is 1.8 at 50 years (i.e., the extrapolated 50-year value is divided by 1.8 to establish the allowable design value).

This testing may be conducted using static pressurization (the standard for water piping) or cyclic pressure testing (which is common for small-diameter pipes used in the oil field industry). The same pipe tested in both static and cyclic pressure conditions will exhibit significantly different regression behavior. The cycling testing condition is far more severe (25 cycles per minute from 0 to test pressure). Because the test is so severe, the common practice is to use the 50-year value directly for design purposes (i.e., the safety factor applied to the extrapolated 50-year value is 1.0). To illustrate the comparison of the two procedures, Figure 2-3 shows the results of a filament-wound epoxy pipe tested both by static and cyclic pressure testing procedures.

This page intentionally blank.

AWWA MANUAL M45

Chapter 3

Manufacturing

3.1 INTRODUCTION

Machine-made fiberglass pipe is produced using two basic processes: filament winding and centrifugal casting. Each process produces a pipe with characteristics that, although unique and advantageous for some applications, will meet the performance requirements of ANSI/AWWA Standard C950, *Fiberglass Pressure Pipe*.

3.2 FILAMENT WINDING

Filament winding is the process of impregnating glass fiber reinforcement with resin, then applying the wetted fibers onto a mandrel in a prescribed pattern. Fillers, if used, are added during the winding process. Chopped glass rovings may be used as supplemental reinforcement. Repeated application of wetted fibers, with or without filler, results in a multilayered structural wall construction of the required thickness. After curing, the pipe may undergo one or more auxiliary operations such as joint preparation. The inside diameter (ID) of the finished pipe is fixed by the mandrel outside diameter (OD). The OD of the finished pipe is variable and determined by the pipe wall thickness.

The filament winding process is illustrated in Figure 3-1. Within the broad definition of filament winding there are several methods used, including reciprocal, continuous, multiple mandrel, and ring and oscillating mandrel, each of which is described briefly. Figure 3-2 shows the application of impregnated glass reinforcement onto a mandrel during production of a filament-wound pipe.

3.2.1 Reciprocal Method

The reciprocal method is the most widely used filament winding production method. In this method the fiber placement head with the associated resin bath drives back and forth past a rotating mandrel (see Figure 3-1). The angle of fiber placement relative to the mandrel axis is controlled by the synchronized translational speed of the bath and the rotational speed of the mandrel.

20 FIBERGLASS PIPE DESIGN

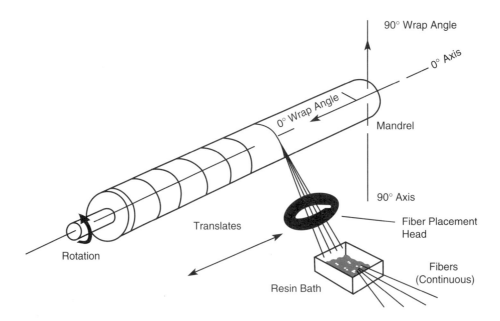

Reprinted with permission from Fiberglass Pipe Handbook, *Fiberglass Pipe Institute, New York, N.Y.*

Figure 3-1 Filament winding process

Figure 3-2 Application of impregnated glass reinforcement of a filament-wound pipe

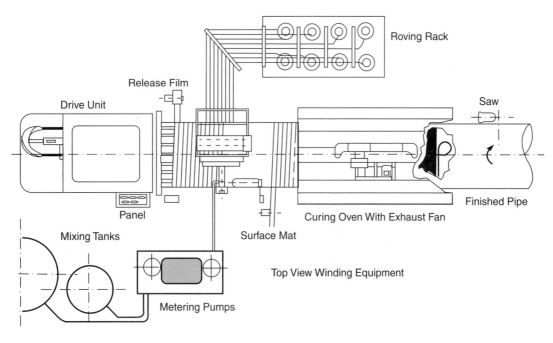

Source: Flowtite Technology, Sandefjord, Norway.

Figure 3-3 Continuous advancing mandrel method

3.2.2 Continuous Methods

In one type of continuous process, pipe is made on one or more mandrels, which move past stations that apply fiberglass tapes preimpregnated with resin or glass fiber and resin. The winding angles are controlled through a combination of longitudinal mandrel speed, mandrel rotation (if used), or the rotation of planetary glass application stations. Once started, these methods produce pipe continuously, stopping only to replenish or change material components.

A second type of continuous process is the continuous advancing mandrel, which is composed of a continuous steel band supported by beams, which form a cylindrically shaped mandrel. The beams rotate, friction pulls the band around, and roller bearings allow the band to move longitudinally so that the entire mandrel continuously moves in a spiral path toward the end of the machine. Raw materials (continuous fibers, chopped fibers, resin, and aggregate fillers) are fed to the mandrel from overhead. Release films and surfacing materials are applied from rolls adjacent to the mandrel. After curing, a synchronized saw unit cuts the pipe to proper length. This method is illustrated in Figure 3-3. Finished pipe emerging from the curing oven is shown in Figure 3-4.

3.2.3 Multiple Mandrel Method

In the multiple mandrel method, a single materials-application system applies wetted glass reinforcement simultaneously to two or more mandrels. When the winding operation finishes, the mandrels are indexed to a new position for curing while another set of mandrels is wound.

3.2.4 Ring and Oscillating Mandrel Method

The use of 360° glass delivery systems, sometimes in combination with an oscillating mandrel, allows production with both high- and low-winding angles as single circuit patterns (without interlayer crossovers).

22   FIBERGLASS PIPE DESIGN

Source: Flowtite Technology, Sandefjord, Norway.

Figure 3-4   Finished pipe emerging from curing oven

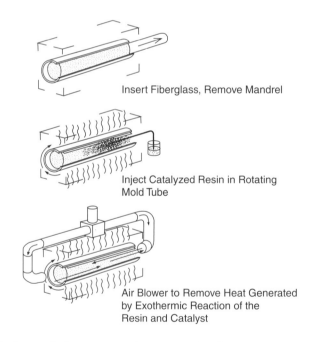

Figure 3-5   Preformed glass reinforcement sleeve method

3.3   CENTRIFUGAL CASTING

Centrifugal casting is used to manufacture tubular goods by applying resin and reinforcement to the inside of a mold that is rotated and heated, subsequently polymerizing (curing) the resin system. The OD of the finished pipe is determined by the ID of the mold tube. The ID of the finished pipe is variable and determined by the amount

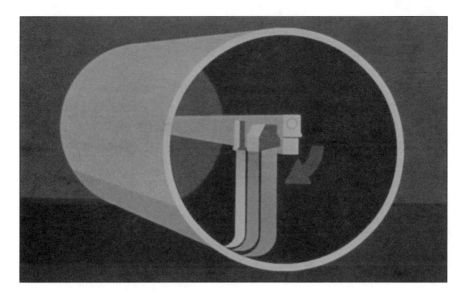

Source: Hobas Pipe USA Inc., Houston, Texas.

Figure 3-6 Chopped glass reinforcement method

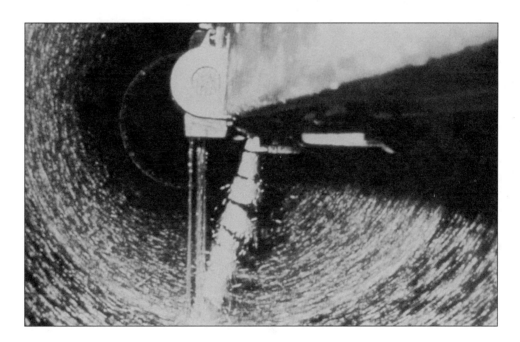

Source: Hobas Pipe USA Inc., Houston, Texas.

Figure 3-7 Application of glass, resin, and sand

of material introduced into the mold. Other materials, such as sand or fillers, may be introduced in the process during manufacture of the pipe.

Two different methods of centrifugal casting are used and are described briefly.

1. *Preformed glass reinforcement sleeve method.* A preformed glass reinforcement sleeve is placed inside a steel mold. As the steel mold rotates, resin and a filler, if used, are placed within the mold by means of a feed tube that

moves in and out of the mold, thus wetting out the preformed sleeve. This method is illustrated in Figure 3-5.

2. *Chopped glass reinforcement method.* Varying proportions of chopped glass reinforcement, resin, and aggregate are introduced simultaneously, by layer, from a feeder arm that moves in and out of the mold. This method is illustrated in Figure 3-6. Application of glass, resin, and sand within a rotating mold is shown in Figure 3-7.

REFERENCE

American Water Works Association. ANSI/AWWA C950, *Standard for Fiberglass Pressure Pipe.* Denver, Colo.: American Water Works Association.

AWWA MANUAL M45

Chapter 4

Hydraulics

4.1 HYDRAULIC CHARACTERISTICS

The smooth interior of fiberglass pipe results in low fluid resistance, which could lower horsepower requirements for pumped systems. Because the interior pipe surface typically remains smooth over time in most fluid services, fluid resistance does not increase with age. In addition, the smooth interior allows the pipe diameter to be reduced while maintaining the desired flow.

This chapter provides a basis for analysis of the flow capacity, economics, and fluid transient characteristics of fiberglass pipe.

4.2 PRELIMINARY PIPE SIZING

The first step in designing a piping system is to determine the pipe size needed to transport a specific amount of fluid. Many engineers have adopted rules that are independent of pipe length but rely on typical or limiting fluid velocities or allowable pressure loss per 100 ft (30 m) of pipe. Once the fluid velocity or the pressure loss is known, it is easy to size a pump to provide the proper flow rate at the required pressure. The following equations are guidelines for the initial sizing of pipe. These equations are presented with inch-pound units in the left-hand column and metric units in the right-hand column.

4.2.1 Maximum Velocity

$$V = 48/\rho^{0.33} \qquad\qquad V = 37/\rho^{0.33} \qquad (4\text{-}1)$$

Where:

V = fluid velocity, ft/sec
ρ = fluid density, lb/ft^3
  = 62.4 lb/ft^3 for water

Where:

V = fluid velocity, m/sec
ρ = fluid density, kg/m^3
  = 1,000 kg/m^3 for water

26 FIBERGLASS PIPE DESIGN

4.2.2 Maximum Velocity for Corrosive or Erosive Fluids

$$V = 24/\rho^{0.33} \qquad\qquad V = 18.4/\rho^{0.33} \qquad (4\text{-}2)$$

4.2.3 Minimum Pipe Diameter for Water

$$d = 0.73[(Q)/(SG)]^{0.5}/\rho^{0.33} \qquad\qquad d = 186[(Q)/(SG)]^{0.5}/\rho^{0.33} \qquad (4\text{-}3)$$

Where:
    d = pipe inside diameter, in.
    Q = flow rate, gpm
    SG = specific gravity, dimensionless
       = 1.0 for water

Where:
    d = pipe inside diameter, mm
    Q = flow rate, L/sec
    SG = specific gravity, dimensionless
       = 1.0 for water

4.2.4 Minimum Pipe Diameter for Corrosive or Erosive Fluids

$$d = 1.03\,[(Q)/(SG)]^{0.5}/\rho^{0.33} \qquad\qquad d = 262[(Q)/(SG)]^{0.5}/\rho^{0.33} \qquad (4\text{-}4)$$

4.3 TYPICAL PIPE DIAMETERS

The equations in Sec. 4.2 represent the minimum pipe diameters or maximum fluid velocities for water and corrosive (or erosive) liquid flow. Typical diameters for fiberglass pressure pipe and suction pipe can be calculated using the following equations.

4.3.1 Typical Diameters for Pressure Pipe Service

$$d = 0.321\,[(Q)/(SG)^2]^{0.434} \qquad\qquad d = 27.1[(Q)/(SG)^2]^{0.434} \qquad (4\text{-}5)$$

4.3.2 Typical Diameters for Suction Pipe Service

$$d = 0.434\,[(Q)/(SG)^2]^{0.434} \qquad\qquad d = 36.6[(Q)/(SG)^2]^{0.434} \qquad (4\text{-}6)$$

4.3.3 Conversion of Flow Rate to Fluid Velocity

$$V = 0.409\,(Q/d^2) \qquad\qquad V = 1{,}274(Q/d^2) \qquad (4\text{-}7)$$

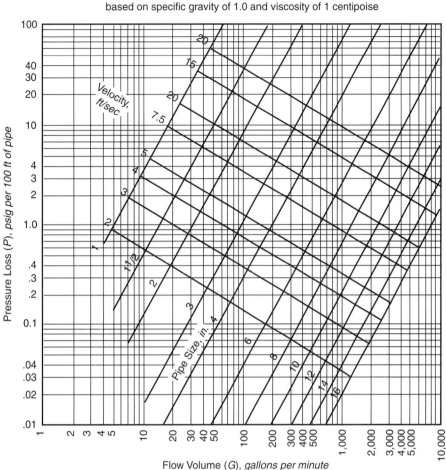

Reprinted with permission from Fiberglass Pipe Handbook, *Fiberglass Pipe Institute,* New York, N.Y.

Figure 4-1 Friction pressure loss due to water flow through fiberglass pipe

4.4 PRESSURE REDUCTION CALCULATIONS

Reduction in pressure, expressed as head loss in feet (meters) or pressure loss pounds in per square inch (kilopascals), occurs in all piping systems because of elevation changes, turbulence caused by abrupt changes of direction, and friction within the pipe and fittings.

Different computational methods can be used to determine the head loss in fiberglass pipe. The most common methods are the Hazen-Williams, Manning, and Darcy-Weisbach equations. The suitability of each method depends on the type of flow (gravity or pumped) and the level of accuracy required.

4.4.1 Hazen-Williams Equation

The Hazen-Williams equation is applicable to water pipes under conditions of full turbulent flow. Although not as technically correct as other methods for all velocities, the Hazen-Williams equation has gained wide acceptance in the water and wastewater industries.

The Hazen-Williams equation is presented in nomograph form in Figure 4-1, which is typical for small-diameter fiberglass pipe. Note, however, that graphical solutions usually are valid for water only. When fluids other than water are encountered, a more universal solution such as the Darcy-Weisbach equation should be used. The Hazen-Williams equation is valid for turbulent flow and will usually provide a conservative solution for determining the head loss in fiberglass pipe.

$$h_f = 0.2083\,(100/C)^{1.85}\,(Q^{1.85}/d^{4.87}) \qquad h_f = 240 \times 10^6\,(100/C)^{1.85}\,(Q^{1.85}/d^{4.87}) \quad (4\text{-}8)$$

Where:
- h_f = friction factor, ft of water/100 ft
- C = Hazen-Williams roughness coefficient, dimensionless (typical value for fiberglass pipe = 150)

Where:
- h_f = friction factor, m of water/100 m
- C = Hazen-Williams roughness coefficient, dimensionless (typical value for fiberglass pipe = 150)

NOTE: Graphs and examples use nominal pipe size for simplicity. The actual inside diameter (ID) should be used in hydraulic calculations.

4.4.2 Simplified Hazen-Williams

Many engineers prefer a simplified version of the Hazen-Williams equation:

$$h_f = [42.7\,Q/(Cd^{2.63})]^{1.852} \qquad h_f = [3.35 \times 10^6\,Q/(Cd^{2.63})]^{1.852} \quad (4\text{-}9)$$

4.4.3 Head Loss Converted to Pressure Loss

Head loss for any liquid is converted into pressure loss using the following equation:

$$p = (H_f)(SG)/2.31 \qquad p = (H_f)(SG)/0.102 \quad (4\text{-}10)$$

Where:
- p = pressure loss, psi
- $H_f = (h_f)(L)/100$, ft
- L = line length, ft

Where:
- p = pressure loss, kPa
- $H_f = (h_f)(L)/100$, m
- L = line length, m

New fiberglass pipe has a Hazen-Williams roughness coefficient C value of 150–165. A design value of 150 is frequently used with fiberglass pipe.

4.4.4 Manning Equation

The Manning equation typically solves gravity flow problems where the pipe is only partially full and is under the influence of an elevation head only.

$$Qm = (1.486/n)(S)^{0.5}(A)(R)^{0.667} \qquad Qm = (1{,}000/n)(S)^{0.5}(A)(R)^{0.667} \quad (4\text{-}11)$$

Where:

Qm = flow rate, ft³/sec
n = Manning roughness coefficient
    = 0.009 for typical fiberglass pipe
S = hydraulic slope, ft/ft = $(H_1 - H_2)/L$
H_1 = upstream elevation, ft
H_2 = downstream elevation, ft
L = length of pipe section, ft
A = cross-sectional area of pipe, ft²
R = hydraulic radius (A/W_p), ft
W_p = wetted perimeter of pipe, ft

Where:

Qm = flow rate, L/sec
n = Manning roughness coefficient
    = 0.009 for typical fiberglass pipe
S = hydraulic slope, m/m = $(H_1 - H_2)/L$
H_1 = upstream elevation, m
H_2 = downstream elevation, m
L = length of pipe section, m
A = cross-sectional area of pipe, m²
R = hydraulic radius (A/W_p), m
W_p = wetted perimeter of pipe, m

4.4.5 Darcy-Weisbach Equation

The Darcy-Weisbach equation states that pressure loss is proportional to the square of the velocity and the length of the pipe. It is inversely proportional to the diameter of the pipe. The primary advantage of this equation is that it is valid for all fluids in both laminar and turbulent flow. The disadvantage is that the Darcy-Weisbach friction factor is a variable. Once preliminary sizing of the pipe diameter has been completed, the next step is to determine whether the flow pattern within the pipe is laminar or turbulent. This characterization of the flow is necessary in the selection of the appropriate friction factor to be used with the Darcy-Weisbach equation. The well-known Reynolds number equation is used to characterize the fluid flow:

$$R_e = (ID)(V)/\mu \qquad R_e = (ID)(V)/\mu \qquad (4\text{-}12)$$

Where:

R_e = Reynolds number, dimensionless
ID = pipe inside diameter, ft
V = fluid velocity, ft/sec
μ = kinematic viscosity, ft²/sec

Where:

R_e = Reynolds number, dimensionless
ID = pipe inside diameter, m
V = fluid velocity, m/sec
μ = kinematic viscosity, m²/sec

This guideline determines the type of flow from the Reynolds number:

Flow Type	Reynolds Number
Laminar flow	$R_e \leq 2{,}000$
Transition flow zone	$2{,}000 < R_e < 4{,}000$
Turbulent flow	$R_e \geq 4{,}000$

Simply stated, the Darcy-Weisbach equation is as follows:

$$H_f = [fL(V^2)]/2(\text{ID})g \qquad H_f = [fL(V^2)]/2(\text{ID})g \qquad (4\text{-}13)$$

Where:

f = Darcy-Weisbach friction factor, dimensionless
g = gravitational constant = 32.2 ft/sec^2

Where:

f = Darcy-Weisbach friction factor, dimensionless
g = gravitational constant = 9.81 m/sec^2

If flow in the pipe is laminar (i.e., $R_e \leq 2{,}000$), the friction factor f_l reduces to

$$f_l = 64/R_e \qquad\qquad f_l = 64/R_e \qquad (4\text{-}14)$$

Where:

f_l = friction factor for laminar flow

Where:

f_l = friction factor for laminar flow

NOTE: Friction factor for laminar flow is denoted as f_l, and f_t denotes friction factor for turbulent flow.

When the flow regime is turbulent (i.e., $R_e \geq 4{,}000$), the friction factor can be determined from the Moody diagram, which is found in most fluid mechanics texts (see Figure 4-2). Fiberglass pipe has a surface roughness parameter e equal to 1.7×10^{-5} ft (5.18×10^{-6} m). When divided by the pipe diameter (e/ID), the friction factor f_t for turbulent flow can be extracted from the smooth pipe segment of the diagram. The friction factor for turbulent flow can also be calculated from the Colebrook equation:

$$1/f_t = -2 \log\,[(e/\text{ID})/3.7] + 2.51/[(R_e)\,(f_t^{0.5})] \quad 1/f_t = -2 \log\,[(e/\text{ID})/3.7] + 2.51/[(R_e)\,(f_t^{0.5})] \quad (4\text{-}15)$$

Where:

f_t = Moody friction factor, dimensionless
e = surface roughness factor, ft
 = 1.7×10^{-5} as typical for fiberglass pipe
ID = pipe inside diameter, ft

Where:

f_t = Moody friction factor, dimensionless
e = surface roughness factor, m
 = 5.18×10^{-6} as typical for fiberglass pipe
ID = pipe inside diameter, m

This equation is difficult to solve because it is implicit in f_t and requires a trial-and-error iterative solution. The following simplified equation relates the friction factor to the Reynolds number and is accurate to within 1% of the Colebrook equation:

$$f_t = [1.8 \log\,(R_e/7)]^{-2} \qquad\qquad f_t = [1.8 \log\,(R_e/7)]^{-2} \qquad (4\text{-}16)$$

4.5 HEAD LOSS IN FITTINGS

Head loss in fittings is frequently expressed as the equivalent length of pipe that is added to the straight run of pipe. This approach has sufficient accuracy for many applications and is used most often with the Hazen-Williams or Manning equations. The approach does not consider turbulence and subsequent losses created by different fluid velocities. When tabular data are not available or when additional accuracy is necessary, head loss in fittings (or valves) can be determined using loss coefficients (K factors) for each type of fitting. Table 4-1 provides the typical K factors. In this

HYDRAULICS 31

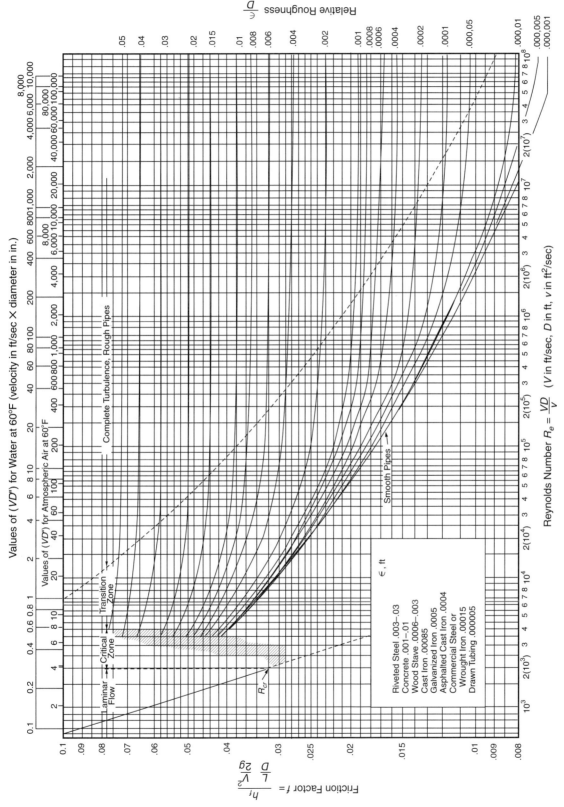

Reprinted with permission from L.F. Moody, Friction Factors for Pipe Flow, ASME, 345 E. 47th St., New York, NY 10017.

Figure 4-2 Moody diagram for determination of friction factor for turbulent flow

32 FIBERGLASS PIPE DESIGN

Table 4-1 Typical *K* factors for fiberglass fittings

Type of Fitting	K Factor
90° elbow, standard	0.5
90° elbow, single miter	1.4
90° elbow, double miter	0.8
90° elbow, triple miter	0.6
180° return bend	1.3
Tee, straight flow	0.4
Tee, flow to branch	1.4
Tee, flow from branch	1.7
Reducer, single size reduction	0.7
Reducer, double size reduction	3.3

approach the *K* factor for each fitting is multiplied by the velocity head of the fluid flow. Equation 4-17 illustrates the loss coefficient approach.

$$H_{ff} = K(V^2/2g) \qquad\qquad H_{ff} = K(V^2/2g) \qquad (4\text{-}17)$$

Where:

H_{ff} = equivalent fittings head loss, ft

K = K factor for each fitting type from Table 4-1

Where:

H_{ff} = equivalent fittings head loss, m

K = K factor for each fitting type from Table 4-1

Many hydraulic handbooks provide *K* factors for various types of fittings and valves not included in this manual.

The total head loss in a system includes, but is not limited to, losses from fittings, the head loss from the straight run pipe, and head losses due to changes in elevation.

4.6 ENERGY CONSUMPTION CALCULATION PROCEDURE

Pipeline operating costs are driven, in large part, by the frictional resistance of the pipe and the corresponding power consumption, and the pipeline design process should consider the operating economics of the pipeline material. This section outlines the basic procedure for determining the head loss due to friction and relative economic merits when considering different pipe materials.

4.6.1 Pipeline Economic Analysis Procedure

Step 1. Calculate the head loss (Eq 4-9):

$$h_f = [42.7\ Q/(Cd^{2.63})]^{1.852} \qquad\qquad h_f = [3.35 \times 10^6\ Q/(Cd^{2.63})]^{1.852}$$

Step 2. Convert head loss to pump horsepower demand:

$$HP = Q\, \rho_1\, H_f/33{,}000 \qquad\qquad HP = Q\, \rho_1\, H_f/102 \qquad (4\text{-}18)$$

Where:
- HP = water pumping power required, hp
- 1 hp = 33,000 ft-lbf/min
- ρ_1 = fluid density = lb/gal
  - = 8.34 lb/gal for water
- $H_f = (h_f)(L)/100$, ft
- L = line length, ft

Where:
- HP = water pumping power required, kW
- 1 kW = 102 kg-m/sec
- ρ_1 = fluid density = kg/L
  - = 1 kg/L for water
- $H_f = (h_f)(L)/100$, m
- L = line length, m

Step 3. Calculate the annual energy usage (To demonstrate the calculations in a clear format, the expressions below assume the pumps run 24 hours per day at full capacity. This is not a realistic assumption. In design situations, engineers must assess the actual expected operating conditions, e.g., 10% of the time at full power, 25% of the time at 75% power, etc.):

$$Ec = (HP)(24)(365)(0.7457)/eff \qquad\qquad Ec = (HP)(24)(365)/eff \qquad (4\text{-}19)$$

Where:
- Ec = annual energy consumption, kW-hr per year
- 24 = hours per day
- 365 = days per year
- 0.7457 = kW-hr/hp-hr
- *eff* = overall (pump and motor) efficiency, typically 75%–85%

Where:
- Ec = annual energy consumption, kW-hr per year
- 24 = hours per day
- 365 = days per year
- *eff* = overall (pump and motor) efficiency, typically 75%–85%

Step 4. Calculate average annual energy cost (AEC):

$$\text{AEC} = (Ec)\,(\text{UEC}) \qquad\qquad \text{AEC} = (Ec)\,(\text{UEC}) \qquad (4\text{-}20)$$

Where:
- AEC = annual energy cost, $
- UEC = unit energy cost, $/kW-hr

Where:
- AEC = annual energy cost, $
- UEC = unit energy cost, $/kW-hr

Techniques that consider the time value of money (net present value, life-cycle costing, etc.) can also be used to evaluate the relative economics of alternative pipe materials. These techniques consider the installed cost of pipe in the calculation and future cash flows are discounted to present value.

4.7 PRESSURE SURGE

Pressure surge, also known commonly as water hammer, results from an abrupt change of fluid velocity within the system. The pressure surge results from the rapidly moving wave that increases and decreases the pressure in the system depending on the source and direction of wave travel. Under certain conditions, pressure surges can reach magnitudes sufficient to rupture or collapse a piping system, regardless of the material of construction.

Rapid valve closure can result in the buildup of pressure waves due to the conversion of kinetic energy of the moving fluid to potential energy that must be accommodated. These pressure waves will travel throughout the piping system and can cause damage far away from the wave source.

4.7.1 Cause and Control of Pressure Surge

The magnitude of pressure surge is a function of the fluid properties and velocity, the modulus of elasticity and wall thickness of the pipe material, the length of the line, and the speed at which the momentum of the fluid changes. The relatively high compliance (low modulus of elasticity) of fiberglass pipe contributes to a self-damping effect as the pressure wave travels through the piping system.

In addition to rapid valve closure or opening, sudden air release and pump start-up or shut-down can create pressure surge. Pressure surges do not show up readily on conventional Bourdon tube gauges because of the slow response of the instrument. The net result of pressure surge can be excessive pressures, pipe vibration, or movement that can cause failure in pipe and fittings.

In some cases, anchoring the piping system may mitigate pipe vibration and movement problems. In other cases, mechanical valve operators, accumulators, rupture discs, surge relief valves, feedback loops around pumps, etc., may be required to protect against or remove the source of pressure surge.

Good design practice usually prevents pressure surge in most systems. Installation of valves that cannot open or close rapidly is one simple precaution. In addition, pumps should never be started in empty discharge lines unless slow-opening, mechanically actuated valves can increase the flow rate gradually.

4.7.2 Pressure Surge Calculation

The Talbot equation calculates surge pressure due to a change in velocity:

$$P_s = (a/g)\,(SG/2.3)\,(\Delta V) \qquad\qquad P_s = (a)\,(SG)\,(\Delta V) \qquad (4\text{-}21)$$

Where: Where:

$$a = 12/[\,(\rho/g)(1/k + d/E\,(t)\,)\,]^{0.5} \qquad a = 1/[\,(\rho/g)(1/10^9 k + d/10^9 E\,(t)\,)\,]^{0.5} \qquad (4\text{-}22)$$

P_s = pressure surge deviation from normal, psig

SG = fluid specific gravity, dimensionless

ΔV = change in flow velocity, ft/sec

P_s = pressure surge deviation from normal, kPa

SG = fluid specific gravity, dimensionless

ΔV = change in flow velocity, m/sec

a = wave velocity, ft/sec
ρ = fluid density, lb/ft^3
g = gravitational constant, 32.2 ft/sec^2
k = bulk modulus of compressibility of liquid, psi
 = 300,000 psi for water
d = pipe inside diameter, in.
E = modulus of elasticity of pipe wall, psi
t = pipe wall thickness, in.

a = wave velocity, m/sec
ρ = fluid density, kg/m^3
g = gravitational constant, 9.81 m/sec^2
k = bulk modulus of compressibility of liquid, GPa
 = 2 GPa for water
d = pipe inside diameter, mm
E = modulus of elasticity of pipe wall, GPa
t = pipe wall thickness, mm

The pressure class P_c must be greater than or equal to the sum of the working pressure P_w and surge pressure P_s divided by 1.4 (see chapter 5, Sec. 5.7.1.3).

Many fluid mechanics and hydraulic handbooks provide procedures such as the previous Talbot equation for calculating pressure surges as a result of a single valve closure in simple piping systems. Sophisticated fluid transient computer programs are also available to analyze pressure surge in complex multibranch piping systems under a variety of conditions.

4.8 DESIGN EXAMPLES

Example 4-1: Use of the Hazen-Williams equation.

Compute the frictional pressure loss in a 1,500-ft long, 10-in. diameter fiberglass pipe transporting 2,000 gpm of water.

(C = 150 as typical)

Compute the frictional pressure loss in a 450-m long, 250-mm diameter fiberglass pipe transporting 130 L/sec of water.

(C = 150 as typical)

Step 1. Compute the head loss per unit length of pipe using Eq 4-9:

$h_f = [42.7 \, Q/(Cd^{2.63})]^{1.852}$
$h_f = [42.7(2{,}000)/(150)(10)^{2.63}]^{1.852}$
 = 1.71 ft water/100 ft

$h_f = [3.35 \times 10^6 \, Q/(Cd^{2.63})]^{1.852}$
$h_f = [3.35 \times 10^6 \, (130)/(150)(250^{2.63})]^{1.852}$
 = 1.95 m water/100 m

The total head loss for the length of pipe in this example is then:

$H_f = 1.71 \, (1{,}500)/100 = 25.65$ ft of water $H_f = 1.95 \, (450)/100 = 8.78$ m of water

Step 2. Convert head loss to pressure drop using Eq 4-10:

$p = (H_f)(SG)/2.31$
 = 25.65(1.0)/2.31 = 11.1 psi

$p = (H_f)(SG)/0.102$
 = 8.78(1.0)/0.102 = 86 kPa

Example 4-2: Determine the pipe diameter, working pressure, and pressure class on a pipeline.

A pipeline requires 5,000 ft of pipe, four 90° elbows (double miter), and three 45° elbows (single miter) with an elevation change of 25 ft. Assume the kinematic viscosity μ = 0.00001. The flow rate is 8,000 gpm.

A pipeline requires 1,500 m of pipe, four 90° elbows (double miter), and three 45° elbows (single miter) with an elevation change of 7.5 m. Assume kinematic viscosity $\mu = 0.93 \times 10^{-6}$. The flow rate is 500 L/sec.

36   FIBERGLASS PIPE DESIGN

Step 1. Determine minimum diameter (Eq 4-3):

$d = 0.73\ [(Q)/(SG)]^{0.5}/\rho^{0.33}$ $d = 186\ [(Q)/(SG)]^{0.5}/\rho^{0.33}$
$d = 0.73\ [(8{,}000)/(1)]^{0.5}/62.4^{0.33}$ $d = 186\ [(500)/(1)]^{0.5}/(1{,}000)^{0.33}$
$\quad = 16.7\ \text{in.}$ $\quad = 425\ \text{mm}$

Therefore, use the next larger available size, which is 18 in. for purposes of this example (ID 18 in. = 1.5 ft).

Therefore, use the next larger available size, which is 450 mm for purposes of this example (ID 450 mm = .45 m).

Step 2. Calculate average fluid velocity (Eq 4-7):

$V = 0.409\ (Q/d^2)$ $V = 1{,}274\ (Q/d^2)$
$\quad = 0.409\ (8{,}000/(18)^2)$ $\quad = 1{,}274\ (500/(450)^2)$
$\quad = 10.1\ \text{ft/sec}$ $\quad = 3.15\ \text{m/sec}$

Step 3. Calculate the Reynolds number (Eq 4-12):

$R_e = (\text{ID})(V)/\mu$ $R_e = (\text{ID})(V)/\mu$
$\quad = (1.5)(10.1)/0.00001$ $\quad = (.45)(3.15)/0.93 \times 10^{-6}$
$\quad = 1.52 \times 10^6$ $\quad = 1.48 \times 10^6$

Because R_e is greater than 4,000, flow is turbulent.

Step 4. Calculate the friction factor (Eq 4-16):

$f_t = [1.8 \log (R_e/7)]^{-2}$ $f_t = [1.8 \log (R_e/7)]^{-2}$
$\quad = [1.8 \log (1.52 \times 10^6/7)]^{-2}$ $\quad = [1.8 \log (1.48 \times 10^6/7)]^{-2}$
$\quad = 0.0108$ $\quad = 0.0109$

Step 5. Calculate system friction loss using Eq 4-13 and Eq 4-17:

Use a factor K for 90° double miter elbows of 0.8 and for 45° single miter elbows of 0.5. Consequently, the total K factor is $4(0.8) + 3(0.5) = 4.7$.

$H_f = K(V^2/2g) + [f_t\ L(V^2)]/(\text{ID})2g$ $H_f = K(V^2/2g) + [f_t\ L(V^2)]/(\text{ID})2g$
$\quad = V^2/2g\ [K + f_t\ L/\text{ID}]$ $\quad = V^2/2g\ [K + f_t\ L/\text{ID}]$
$\quad = (10.1)^2/2(32.2)\ [4.7 +$ $\quad = (3.15)^2/2(9.81)\ [4.7 +$
$\qquad [0.0108\ (5{,}000)]/(1.5)]$ $\qquad [0.0109\ (1{,}500)]/(.45)]$
$\quad = 64.5\ \text{ft}$ $\quad = 20.8\ \text{m}$

Step 6. Combine friction and elevation head:

$H_{\text{total}} = 64.5 + 25 = 89.5\ \text{ft}$ $H_{\text{total}} = 20.8 + 7.5 = 28.3\ \text{m}$

Step 7. Convert head loss to working pressure (Eq 4-10):

$p = (H_f)(\text{SG})/2.31$ $p = (H_f)(\text{SG})/0.102$
$\quad = (89.5)(1.0)/2.31 = 38.7\ \text{psi}$ $\quad = (28.3)(1.0)/0.102 = 277\ \text{kPa}$

HYDRAULICS 37

The total working pressure of 38.7 psi would require a 50-psi pressure class. However, a higher pressure class may tentatively be selected to account for possible water hammer in the line. For these examples assume that a 100-psi class is selected. See example 4-4 to verify that this is adequate for pressure surge.	The total working pressure of 277 kPa would require a 350-kPa pressure class. However, a higher pressure class may tentatively be selected to account for possible water hammer in the line. For these examples assume that a 700-kPa class is selected. See example 4-4 to verify that this is adequate for pressure surge.

Example 4-3: Comparative power cost calculation.

Assume a 10,000-ft long, 6-in. diameter pipeline is to deliver 500 gpm of water on a year-round basis. The engineer is considering using fiberglass pipe with an average Hazen-Williams coefficient $C = 150$ and another material that will have an average Hazen-Williams coefficient $C = 100$ over the life of the pipeline. Calculate the average AEC for each candidate material and the total energy cost over a 20-year service life using a unit cost of power of \$0.06/kW-hr.	Assume a 3,000-m long, 150-mm diameter pipeline is to deliver 30 L/sec of water on a year-round basis. The engineer is considering using fiberglass pipe with an average Hazen-Williams coefficient $C = 150$ and another material that will have an average Hazen-Williams coefficient $C = 100$ over the life of the pipeline. Calculate the average AEC for each candidate material and the total energy cost over a 20-year service life using a unit cost of power of \$0.06/kW-hr.

Step 1. Calculate the head loss for each material (Eq 4-9):

$h_f = [42.7 \, Q/(Cd^{2.63})]^{1.852}$ $\qquad\qquad h_f = [3.35 \times 10^6 \, Q/(Cd^{2.63})]^{1.852}$

For fiberglass pipe:

$h_f = [42.7 \times 500/(150 \times 6^{2.63})]^{1.852}$ $\qquad h_f = [3.35 \times 10^6 \times 30/(150 \times 150^{2.63})]^{1.852}$

$= 1.58$ ft/100 ft $\qquad\qquad\qquad\qquad\qquad = 1.55$ m/100 m

$H_f = 158$ ft for the entire line $\qquad\qquad\;\; H_f = 46.5$ m for the entire line

For alternate material:

$= [42.7 \times 500/(100 \times 6^{2.63})]^{1.852}$ $\qquad\;\; = [3.35 \times 10^6 \times 30/(100 \times 150^{2.63})]^{1.852}$

$= 3.34$ ft/100 ft $\qquad\qquad\qquad\qquad\qquad = 3.29$ m/100 m

$= 334$ ft for the entire line $\qquad\qquad\qquad = 98.6$ m for the entire line

Step 2. Convert head loss to horsepower demand (Eq 4-18):

$HP = Q \, \rho_1 \, H_f/33{,}000$ $\qquad\qquad\qquad HP = Q \, \rho_1 \, H_f/102$

For fiberglass pipe:

$\qquad = 500(8.34)(158)/33{,}000$ $\qquad\qquad\quad = 30(1)(46.5)/102$

$\qquad = 19.96$ hp $\qquad\qquad\qquad\qquad\qquad\;\; = 13.68$ kW

For alternate material:

= 500(8.34)(334)/33,000   = 30(1)(98.6)/102
= 42.20 hp   = 29 kW

Step 3. Calculate the annual energy consumption *Ec* using an 80% overall pump efficiency and Eq 4-19 (as noted before, for simplicity in demonstrating the calculation procedure, this example assumes constant pump operation at full power. In design situations, engineers must assess actual operating levels.):

$Ec = (HP)(24)(365)(0.7457)/eff$ $Ec = (HP)(24)(365)/eff$

For fiberglass pipe:

= $(HP)(24)(365)(0.7457)/eff$ = $(HP)(24)(365)/eff$
= 19.96(24)(365)(0.7457)/0.80   = 13.68(24)(365)/0.80
= 163,000 kW-hr   = 150,000 kW-hr

For alternate material:

= $(HP)(24)(365)(0.7457)/eff$ = $(HP)(24)(365)/eff$
= 42.20(24)(365)(0.7457)/0.80   = 29(24)(365)/0.80
= 344,620 kW-hr   = 318,000 kW-hr

Step 4. Calculate the AEC (Eq 4-20) and calculate the total energy cost over 20 years:

AEC = (*Ec*)(UEC)   AEC = (*Ec*)(UEC)

For fiberglass pipe:

= 163,000 (0.06) = $9,780/yr   = 150,000 (0.06) = $9,000/yr
= $195,600 over 20 years   = $180,000 over 20 years

For alternate material:

= 344,620 (0.06) = $20,676/yr   = 318,000 (0.06) = $19,080/yr
= $413,000 over 20 years   = $381,600 over 20 years

Example 4-4: Surge pressure calculation.

Determine if the maximum pressure surge for the pipe in example 4-2 is within the 40% surge allowance criteria. Assume a full instantaneous change in velocity equal to the flow velocity in the pipe. The fiberglass pipe has a tensile modulus of 3×10^6 psi and a pressure class of 100 psi. The pipe wall thickness $t = 0.21$ in. The bulk modulus of water is 300,000 psi.

Determine if the maximum pressure surge for the pipe in example 4-2 is within the 40% surge allowance criteria. Assume a full instantaneous change in velocity equal to the flow velocity in the pipe. The fiberglass pipe has a tensile modulus of 20 GPa and a pressure class of 700 kPa. The pipe wall thickness $t = 5.3$ mm. The bulk modulus of water is 2.07 GPa.

Step 1. Calculate the wave velocity (Eq 4-22):

$a = 12/[(\rho/g)(1/k + d/E(t))]^{0.5}$
$= 12/[(62.4/32.2)(1/3 \times 10^5$
$+ 18/3 \times 10^6 (0.21))]^{0.5}$
$= 1{,}526$ ft/sec

$a = 1/[\rho/g (1/10^9 k + d/10^9 E(t))]^{0.5}$
$= 1/[1{,}000(9.81)/9.81(1/10^9(2)$
$+ 450/10^9(20)(5.3))]^{0.5}$
$= 459$ m/sec

Step 2. Calculate the surge pressure (Eq 4-21):

$P_s = (a/g)(SG/2.3)(\Delta V)$
$= (1{,}526/32.2)(1/2.3)(10.1)$
$= 208$ psi

$P_s = (a)(SG)(\Delta V)$
$= (459)(1)(3.15)$
$= 1{,}446$ kPa

Step 3. Check compliance with the maximum system pressure requirement:

$P_c \geq (P_w + P_s)/1.40$

$P_c \geq (P_w + P_s)/1.4$

From example 4-2, $P_w = 38.7$ psi and the selected pressure class was $P_c = 100$ psi:

$(38.7 + 208)/1.4 = 176$ psi

From example 4-2, $P_w = 277$ kPa and the selected pressure class was $P_c = 700$ kPa:

$(277 + 1{,}446)/1.4 = 1{,}231$ kPa

This exceeds the pressure class. The engineer has three options. The first would be to increase the pressure class to accommodate the surge, maintaining the same pipe diameter. The second would be to increase pipe diameter, which together with a more moderate increase in pressure class would satisfy the maximum system pressure requirement. The larger pipe diameter will lower operating pressure due to lower friction loss and will lower fluid velocity. The third option is to provide measures, such as a surge tank, to reduce the magnitude of the surge.

This exceeds the pressure class. The engineer has three options. The first would be to increase the pressure class to accommodate the surge, maintaining the same pipe diameter. The second would be to increase pipe diameter, which together with a more moderate increase in pressure class would satisfy the maximum system pressure requirement. The larger pipe diameter will lower operating pressure due to lower friction loss and will lower fluid velocity. The third option is to provide measures, such as a surge tank, to reduce the magnitude of the surge.

For this example, the second option will be used and a diameter of 20 in. (wall thickness of 0.23 in.) with a pressure class of 150 psi will be evaluated.

For this example, the second option will be used and a diameter of 500 mm (wall thickness of 5.8 mm) with a pressure class of 1,000 kPa will be evaluated.

Step 4. Calculate the fluid velocity for the new pipe diameter (Eq 4-7):

$V = 0.409(Q/d^2)$
$= 0.409[8{,}000/(20)^2]$
$= 8.18$ ft/sec

$V = 1{,}274(Q/d^2)$
$= 1{,}274[500/(500)^2]$
$= 2.55$ m/sec

Note that this velocity is lower than the 10.1 ft/sec (3.15 m/sec) in example 4-2.

Step 5. Calculate the new working pressure.

A. Reynolds number (Eq 4-12):

$R_e = (ID)(V)/\mu$
$= (20/12)(8.18)/0.00001$
$= 1,363,333$

$R_e = (ID)(V)/\mu$
$= (500/1,000)(2.55)/0.93 \times 10^{-6}$
$= 1,370,967$

B. Friction factor (Eq 4-16):

$f_t = [1.8 \log(R_e/7)]^{-2}$
$= [1.8 \log(1,363,333/7)]^{-2}$
$= 0.01103$

$f_t = [1.8 \log(R_e/7)]^{-2}$
$= [1.8 \log(1,370,967/7)]^{-2}$
$= 0.01102$

C. Friction losses using Eq 4-13 and Eq 4-17:

Use a factor K for 90° double miter elbows of 0.8 and for 45° single miter elbows of 0.5. The total K factor is then $4(0.8) + 3(0.5) = 4.7$.

$H_f = K(V^2/2g) + [f_t L(V^2)]/2(ID)g$
$= V^2/2g [K + f_t L/ID]$
$= (8.18)^2/2(32.2) [4.7 +$
$[0.01103(5,000)]/(20/12)]$
$= 39.3$ ft

$H_f = K(V^2/2g) + [f_t L(V^2)]/2(ID)g$
$= V^2/2g [K + f_t L/ID]$
$= (2.55)^2/2(9.81) [4.7 +$
$[0.01102(1,500)]/.5]$
$= 12.5$ m

Combine friction and elevation head:

$H_{\text{total}} = 39.3 + 25 = 64.3$ ft

$H_{\text{total}} = 12.5 + 7.5 = 20$ m

D. Convert to working pressure (Eq 4-10) and using H_{total} for H_f:

$p = (H_f)(SG)/2.31$
$= (64.3)(1.0)/2.31 = 27.8$ psi

$p = (H_f)(SG)/0.102$
$= (20)(1.0)/0.102 = 196$ kPa

Step 6. Calculate the wave velocity (Eq 4-22):

$a = 12/[(\rho/g)(1/k + d/E(t))]^{0.5}$
$= 12/[(62.4/32.2)(1/3 \times 10^5 +$
$20/3 \times 10^6 (0.23))]^{0.5}$
$= 1,516$ ft/sec

$a = 1/[(\rho/g)(1/10^9 k + d/10^9 E(t))]^{0.5}$
$= 1/[(1,000(9.81)/9.81(1/10^9(2) +$
$500/10^9(20)(5.8))]^{0.5}$
$= 456$ m/sec

Step 7. Calculate the pressure surge using (Eq 4-21):

$P_s = (a/g)(SG/2.3)(\Delta V)$
$= (1,516/32.2)(1/2.3)(8.18)$
$= 167$ psi

$P_s = (a)(SG)(\Delta V)$
$= (456)(1)(2.55)$
$= 1,162$ kPa

Check compliance with maximum system pressure requirement:

$P_c \geq (P_w + P_s) / 1.4$ $P_c \geq (P_w + P_s) / 1.4$
$150 \geq (27.8 + 167)/1.4$ $1{,}000 \text{ kPa} \geq (196 + 1{,}162)/1.4$
$150 \geq 139$ psi $1{,}000 \text{ kPa} \geq 970 \text{ kPa}$

Therefore, the system pressure requirement is satisfied by using the higher pressure class in a larger diameter. Before final selection, the engineer would typically evaluate the economics of using the larger diameter with a higher pressure class versus using the original diameter with a still higher pressure class.

REFERENCES

American Water Works Association. ANSI/AWWA C950, *Standard for Fiberglass Pressure Pipe*. Denver, Colo.: American Water Works Association.

Benedict, R.P. 1980. *Fundamentals of Pipe Flow*. New York: John Wiley & Sons.

Brater, E.F., and H.W. King. 1982. *Handbook of Hydraulics*. 6th ed. New York: McGraw-Hill.

Fiberglass Pipe Institute. 1989. *Fiberglass Pipe Handbook*. New York: Fiberglass Pipe Institute.

Kent, G.R. 1978. Preliminary Pipeline Sizing. *Chemical Engineering*.

Sharp, W.W., and T.M. Walski. 1988. Predicting Internal Roughness in Water Mains. *Jour. AWWA*, 80(11):34.

This page intentionally blank.

AWWA MANUAL M45

Chapter 5

Buried Pipe Design

5.1 INTRODUCTION

The structural design procedure for buried fiberglass pipe involves establishing design conditions, selecting pipe classes and corresponding pipe properties, selecting installation parameters, and performing pertinent calculations to ensure that the design requirements of Sec. 5.7 are satisfied. If the results of any calculation indicate that a requirement is not satisfied, it will be necessary to upgrade installation parameters or select a pipe with different properties, or both, and redo pertinent calculations. Special information and calculations not covered in this chapter may be required in unusual cases (see Sec. 5.9).

Both rigorous and empirical methods are used to design fiberglass pipe. In addition to short-term tests, many performance limits are determined at 50 years through statistical extrapolation of data obtained from long-term tests under simulated service conditions. Design stress or strain values are obtained by reducing performance limits using appropriate design factors. Design factors are established to ensure adequate performance over the intended service life of the pipe by providing for variations in material properties and loads not anticipated by design calculations. Design factors are based on judgment, past experience, and sound engineering principles.

The design method discussed in this chapter applies in concept to pipe with uniform walls and to pipe with ribbed-wall cross sections. However, for design of pipe with ribbed walls, some of the equations must be modified to allow for the special properties of this pipe. Also, additional calculations not addressed in this chapter may be required to ensure an adequate design for a ribbed-wall cross section. The equations are presented with inch-pound units in the left-hand column and metric units in the right-hand column.

5.2 TERMINOLOGY

5.2.1 Definitions

The following definitions apply to buried pipe design as discussed in this chapter:

Working pressure, P_w. The maximum anticipated, long-term operating pressure of the fluid system resulting from typical system operation.

Pressure class, P_c. The maximum sustained pressure for which the pipe is designed in the absence of other loading conditions.

Surge pressure, P_s. The pressure increase above the working pressure, sometimes called water hammer, that is anticipated in a system as a result of a change in the velocity of the fluid, such as when valves are operated or when pumps are started or stopped.

Surge allowance, P_{sa}. That portion of the surge pressure that can be accommodated without changing pressure class. The surge allowance is expected to accommodate pressure surges usually encountered in typical systems.

Hydrostatic design basis, HDB. The long-term hydrostatic hoop strength of a specific fiberglass pipe material as determined by tests and detailed evaluation procedures in accordance with ANSI/AWWA Standard C950, pressure classes subsection on long-term hydrostatic design pressure.

Design factor, FS. A specific number greater than 1 used to reduce a specific mechanical or physical property in order to establish a design value for use in calculations.

5.2.2 Symbols

B_d = trench width at pipe springline, in. (mm)
C_n = scalar calibration factor (used in buckling Eq 5-24)
D = mean pipe diameter, in. (mm)
D_f = shape factor per Sec. 5.7.2 (dimensionless)
D_L = deflection lag factor (dimensionless)
E = ring flexural modulus of elasticity, psi (GPa)
E_H = hoop tensile modulus of elasticity, psi (GPa)
EI = stiffness factor per unit length of pipe wall, in.^{2}-lb/in. (m^2-N/m)
F = load per unit length, lb/in. (N/m)
FS = design factor (dimensionless)
$F/\Delta y$ = pipe stiffness, psi (kPa)
H = burial depth to top of pipe, ft (m)
h = height of ground surface above top of pipe, in. (m)
h_{int} = depth at which load from wheels interact, in. (m)
h_w = height of water surface above top of pipe, in. (m)
HDB = hydrostatic design basis, psi (MPa) (for stress basis) or in./in. (mm/mm) (for strain basis)
ID = inside diameter, in. (mm)
I = moment of inertia of pipe wall for ring bending, in.^{4}/in. (mm^4/mm)
I_f = impact factor (dimensionless)
K_x = bedding coefficient (dimensionless)
k_υ = modulus correction factor for Poisson's ratio υ of the soil (dimensionless)
L_1 = dimension of area of wheel load at pipe crown depth in the direction of travel, in. (m) (see Sec. 5.7.3.6)
L_2 = dimension of area of wheel load at pipe crown depth transverse to the direction of travel, in. (m) (see Sec. 5.7.3.6)
M_p = multiple presence factor (dimensionless)
M_s = composite soil constrained modulus, psi (MPa)
M_{sb} = constrained modulus of the pipe zone embedment, psi (MPa)

M_{sn} = constrained modulus of the native soil at pipe elevation, psi (MPa)
OD = outside diameter, in. (mm)
P = vehicular traffic load (wheel load), lb (N)
PS = pipe stiffness, psi (kPa)
P_c = pressure class, psi (kPa)
P_s = surge pressure, psi (kPa)
P_v = internal vacuum pressure, psi (kPa)
P_w = working pressure, psi (kPa)
q_a = allowable buckling pressure, psi (kPa)
q_u = unconfined compressive strength, US tons/ft^2 (kPa)
r = mean pipe radius, in. (mm)
r_c = rerounding coefficient (dimensionless)
R_h = buckling strength correction factor for depth of fill (dimensionless)
R_w = water buoyancy factor (dimensionless)
S_b = long-term, ring-bending strain, in./in. (mm/mm)
S_c = soil support combining factor (dimensionless)
S_i = ultimate hoop tensile strength, psi (MPa)
S_r = hoop tensile stress, psi (MPa), or strain, in./in. (mm/mm), at pressure class
t = thickness of pipe reinforced wall, per ASTM D3567, in. (mm)
t_L = thickness of liner (when used), in. (mm)
t_l = length of tire footprint, in. (mm)
t_w = width of tire footprint, in. (mm)
t_t = total thickness of pipe wall and liner (when used), in. (mm)
W_c = vertical soil load on the pipe, psi (N/m^2)
W_L = live load on the pipe, psi (N/m^2)
γ_s = specific weight of the soil, lb/ft^3 (N/m^3)
γ_w = specific weight of water, lb/in.^{3} (N/m^3)
υ = Poisson's ratio of soil (dimensionless)
Δy = predicted vertical pipe deflection, in. (mm)
Δy_a = maximum allowable long-term vertical pipe deflection, in. (mm)
Δy_t = vertical pipe deflection—in. (mm) when tested by ASTM D2412 with a vertical diameter reduction of 5%
σ_b = maximum ring-bending stress due to deflection, psi (MPa)
σ_c = maximum stress due to combined loading, psi (MPa)
σ_{pr} = working stress due to internal pressure, psi (MPa)
δ_d = maximum permitted long-term installed deflection, in. (mm)
φ_s = factor to account for variability in stiffness of compacted soil (dimensionless)
ε_b = maximum ring-bending strain due to deflection, in./in. (mm/mm)
ε_c = maximum strain due to combined loading, in./in. (mm/mm)
ε_{pr} = working strain due to internal pressure, in./in. (mm/mm)

46   FIBERGLASS PIPE DESIGN

5.3   DESIGN CONDITIONS

Design conditions are largely determined by required flow rate and flow velocity limitations, hydraulics, pipeline elevations and associated geology and topography, available rights-of-way, and installation requirements.

5.3.1   Head Losses

Hydraulic head loss due to pipe friction may be significantly lower for fiberglass pipe than for other types of pipe due to fiberglass pipe's generally smoother bores and freedom from tuberculation and corrosion. This is reflected in typical long-term flow coefficient values of 0.009 for Manning's n and 150 for the Hazen-Williams' C. The engineer may wish to consider this in establishing design conditions. (See chapter 4 on hydraulics.)

5.3.2   Surge Pressures

Surge pressures should be calculated on the basis of the pipe hoop tensile modulus and thickness-to-diameter ratio for given system design parameters (discussed later in this chapter). Excessive surge pressures should be identified in the design phase, and the causative condition should be eliminated or automatic surge-pressure relief provided, otherwise, a higher pressure class should be selected.

5.3.3   Basic Design Conditions

Design conditions that should be established before performing structural design calculations are as follows:

- nominal pipe size (tables 1 through 6 of ANSI/AWWA Standard C950),
- working pressure, P_w,
- surge pressure, P_s,
- soil conditions for the pipe zone embedment and native material at pipe depth,
- soil specific weight, γ_s,
- depth of cover, minimum and maximum,
- vehicular traffic load, P,
- internal vacuum pressure, P_v, and
- average and maximum service temperature.

5.4   PIPE PROPERTIES

Preliminary pipe pressure class selection can usually be made on the basis of working pressure, surge pressure, and external loads, established in Sec. 5.7. Properties at the anticipated average and maximum service temperature for a given class of a specific pipe product should be obtained from the manufacturer or the manufacturer's literature. Values for ring stiffness, axial strength, and hoop tensile strength given in ANSI/AWWA Standard C950 are minimum requirements. Some pipe products may have significantly higher values for these properties. The design may require material properties and structural capacities greater than those given as minimums in ANSI/AWWA Standard C950. Pipe properties necessary for design calculations include the following:

- nominal reinforced wall thickness t and liner thickness t_L (ANSI/AWWA Standard C950),
- hoop tensile modulus of elasticity, E_H,
- hydrostatic design basis, HDB,
- ring flexural modulus of elasticity, E,
- minimum pipe stiffness, $F/\Delta y$ (ANSI/AWWA Standard C950), and
- long-term ring-bending strain, S_b.

5.5 INSTALLATION PARAMETERS

The primary installation parameters that must be selected according to the site conditions and planned installation are the type of backfill soil immediately around the pipe (pipe zone backfill), degree of compaction, and the characteristics of the native soil at the pipe elevation. Initial selection of these parameters may be controlled by prevailing standard specifications, the project soil's boring report, manufacturers' recommendations, or past experience. A given combination of soil type and degree of compaction will largely determine the following values required for design calculations:

- bedding coefficient, K_x (Sec. 5.7.3.4),
- constrained soil moduli for the native soil (M_{sn}) and for the pipe zone embedment (M_{sb}) (Sec. 5.7.3.8), and
- deflection lag factor, D_L (Sec. 5.7.3.3).

5.6 DESIGN PROCEDURE

With conditions, properties, and installation parameters established in accordance with Sec. 5.3 through Sec. 5.5, satisfaction of the requirements listed in Sec. 5.7 can be checked by design calculations. The calculations may be made using either stress or strain, depending on the basis used to establish a particular product performance limit. The procedure for using design calculations to determine whether pipe meets the requirements discussed in Sec. 5.7 is as follows:

1. Calculate P_c from HDB and pipe dimensions (Sec. 5.7.1.1).
2. Check working pressure, P_w (Sec. 5.7.1.2).
3. Check surge pressure, P_s (Sec. 5.7.1.3).
4. Calculate allowable deflection from ring bending (Sec. 5.7.2).
5. Determine soil loads, W_c, and live loads, W_L (Sec. 5.7.3.5 and Sec. 5.7.3.6, respectively).
6. Calculate the composite constrained soil modulus, M_s (Sec. 5.7.3.8).
7. Check deflection prediction, $\Delta y/D$ (Sec. 5.7.3).
8. Check combined loading (Sec. 5.7.4).
9. Check buckling (Sec. 5.7.5).

See Sec. 5.10 for step-by-step example design calculations.

5.7 DESIGN CALCULATIONS AND REQUIREMENTS

5.7.1 Internal Pressure

5.7.1.1 Pressure class, P_c. The pressure class in ANSI/AWWA Standard C950 is related to the long-term strength, or HDB, of the pipe as follows:

For stress basis HDB:

$$P_c < \left(\frac{\text{HDB}}{FS}\right)\left(\frac{2t}{D}\right) \qquad P_c < \left(\frac{\text{HDB}}{FS}\right)\left(\frac{2t}{D}\right) \times 10^3 \qquad (5\text{-}1)$$

For strain basis HDB:

$$P_c < \left(\frac{\text{HDB}}{FS}\right)\left(\frac{2tE_H}{D}\right) \qquad P_c < \left(\frac{\text{HDB}}{FS}\right)\left(\frac{2tE_H}{D}\right) \times 10^6 \qquad (5\text{-}2)$$

Where:
    P_c = pressure class, psi
    HDB = hydrostatic design basis, psi, for stress basis, or in./in. for strain basis
    FS = minimum design factor, 1.8
    t = pipe reinforced wall thickness, in.
    D = mean pipe diameter, in.

Where:
    P_c = pressure class, kPa
    HDB = hydrostatic design basis, MPa, for stress basis, or mm/mm for strain basis
    FS = minimum design factor, 1.8
    t = pipe reinforced wall thickness, mm
    D = mean pipe diameter, mm

$$D = \text{ID} + 2t_L + t; \text{ or} \qquad D = \text{ID} + 2t_L + t; \text{ or}$$

$$D = \text{OD} - t \qquad D = \text{OD} - t$$

Where:
    t_L = liner thickness, in.
    ID = inside diameter, in.
    OD = outside diameter, in.
    E_H = hoop tensile modulus of elasticity for pipe, psi

Where:
    t_L = liner thickness, mm
    ID = inside diameter, mm
    OD = outside diameter, mm
    E_H = hoop tensile modulus of elasticity for pipe, GPa

Hydrostatic design basis (HDB). The HDB of fiberglass pipe varies for different products, depending on the materials and composition used in the reinforced wall and in the liner. The HDB may be defined in terms of reinforced wall hoop stress or hoop strain on the inside surface.

Temperature and service life. The HDB at ambient temperature must be established by testing in accordance with ANSI/AWWA Standard C950 for each fiberglass pipe product by each manufacturer. The required practice is to define projected product performance limits at 50 years. Performance limits at elevated temperature depend on the materials and type of pipe wall construction used. The manufacturer should be consulted for HDB values appropriate for elevated temperature service.

Design factors. Two separate design factors are required in ANSI/AWWA Standard C950 for internal pressure design.

BURIED PIPE DESIGN   49

The first design factor is the ratio of short-term ultimate hoop tensile strength S_i to hoop tensile stress S_r at pressure class P_c. This factor ensures that the stress or strain due to the short-term peak pressure conditions do not exceed the short-term hydrostatic strength of the pipe. The hoop tensile strength values given in Table 10 of ANSI/AWWA Standard C950 reflect a minimum design factor of 4.0 on initial hydrostatic strength.

The second design factor is the ratio of HDB to hoop stress or strain S_r at pressure class P_c. This factor ensures that stress or strain due to sustained working pressure does not exceed the long-term hoop strength of the pipe as defined by HDB. For fiberglass pipe design, this minimum design factor is 1.8.

Both design factors should be checked. Either design factor may govern pipe design, depending on long-term strength regression characteristics of the particular pipe product. Prudent design practice may dictate an increase or decrease in either design factor, depending on the certainty of the known service conditions.

5.7.1.2 Working pressure, P_w. The pressure class of the pipe should be equal to or greater than the working pressure in the system, as follows:

$$P_c \geq P_w \qquad\qquad P_c \geq P_w \qquad (5\text{-}3)$$

Where:

P_w = working pressure, psi

Where:

P_w = working pressure, kPa

5.7.1.3 Surge pressure, P_s. The pressure class of the pipe should be equal to or greater than the maximum pressure in the system, due to working pressure plus surge pressure, divided by 1.4, as follows:

$$P_c \geq \frac{(P_w + P_s)}{1.4} \qquad\qquad P_c \geq \frac{(P_w + P_s)}{1.4} \qquad (5\text{-}4)$$

Where:

P_s = surge pressure, psi

Where:

P_s = surge pressure, kPa

The treatment of surge pressures reflects the characteristics of the pipe and materials covered by ANSI/AWWA Standard C950. Factory hydrotesting at pressures up to $2 P_c$ is acceptable and is not governed by Eq 5-3 and Eq 5-4.

Calculated surge pressure, P_s. The surge pressure calculations should be performed using recognized and accepted theories. (See chapter 4 on hydraulics.)

Calculated surge pressure magnitudes are highly dependent on the hoop tensile elastic modulus and thickness-to-diameter (t/D) ratio of the pipe. Because of this, the engineer should generally expect lower calculated surge pressures for fiberglass pipe than for pipe materials with a higher modulus or thicker wall or both. For example, an instantaneous change in flow velocity of 2 ft/sec (0.6 m/sec) would result in a calculated surge pressure increase of approximately 40 psi (276 kPa) for fiberglass pipe with a modulus of 3,000,000 psi (20.7 GPa) and a t/D ratio of 0.01.

Surge allowance. The surge allowance is intended to provide for rapid transient pressure increases typically encountered in transmission systems. The surge pressure allowance of 0.4 P_c is based on the increased strength of fiberglass pipe for rapid strain rates. Special consideration should be given to the design of systems subject to rapid and frequent cyclic service. The manufacturer should be consulted for specific recommendations.

5.7.2 Ring Bending

The maximum allowable long-term vertical pipe deflection should not result in a ring-bending strain (or stress) that exceeds the long-term, ring-bending strain capability of the pipe reduced by an appropriate design factor. Satisfaction of this requirement is assured by using one of the following formulas.

For stress basis:

$$\sigma_b = D_f E \left(\frac{\Delta y_a}{D}\right)\left(\frac{t_t}{D}\right) \leq \frac{S_b E}{FS} \qquad \sigma_b = 10^3 D_f E \left(\frac{\Delta y_a}{D}\right)\left(\frac{t_t}{D}\right) \leq 10^3 \frac{S_b E}{FS} \quad (5\text{-}5)$$

For strain basis:

$$\varepsilon_b = D_f \left(\frac{\Delta y_a}{D}\right)\left(\frac{t_t}{D}\right) \leq \frac{S_b}{FS} \qquad \varepsilon_b = D_f \left(\frac{\Delta y_a}{D}\right)\left(\frac{t_t}{D}\right) \leq \frac{S_b}{FS} \quad (5\text{-}6)$$

Where:

σ_b = maximum ring-bending stress due to deflection, psi

D_f = shape factor per Table 5-1, dimensionless

E = ring flexural modulus of elasticity for the pipe, psi

Δy_a = maximum allowable long-term vertical pipe deflection, in.

S_b = long-term, ring-bending strain for the pipe, in./in.

D = mean pipe diameter, in.

FS = design factor, 1.5

ε_b = maximum ring-bending strain due to deflection, in./in.

t_t = total wall thickness, in.

$t_t = t + t_L$

Where:

σ_b = maximum ring-bending stress due to deflection, MPa

D_f = shape factor per Table 5-1, dimensionless

E = ring flexural modulus of elasticity for the pipe, GPa

Δy_a = maximum allowable long-term vertical pipe deflection, mm

S_b = long-term, ring-bending strain for the pipe, mm/mm

D = mean pipe diameter, mm

FS = design factor, 1.5

ε_b = maximum ring-bending strain due to deflection, mm/mm

t_t = total wall thickness, mm

$t_t = t + t_L$

5.7.2.1 Shape factor, D_f. The shape factor relates pipe deflection to bending stress or strain and is a function of pipe stiffness, pipe zone embedment material and compaction, haunching, native soil conditions, and level of deflection. Table 5-1 gives values for D_f, assuming inconsistent haunching, deflections of at least 2 to 3%, and stable native soils or adjustments to trench width to offset poor conditions. Values

Table 5-1 Shape factors

		Pipe-Zone Embedment Material and Compaction			
		Gravel*		Sand†	
Pipe Stiffness		Dumped to Slight‡	Moderate to High§	Dumped to Slight‡	Moderate to High§
psi	kPa	Shape Factor, D_f (dimensionless)			
9	62	5.5	7.0	6.0	8.0
18	124	4.5	5.5	5.0	6.5
36	248	3.8	4.5	4.0	5.5
72	496	3.3	3.8	3.5	4.5

* GW, GP, GW–GC, GW–GM, GP–GC, and GP–GM per ASTM D2487 (includes crushed rock).
† SW, SP, SM, SC, GM, and GC or mixtures per ASTM D2487.
‡ <85% Proctor density (ASTM D698), <40% relative density (ASTM D4253 and D4254).
§ ≥85% Proctor density (ASTM D698), ≥40% relative density (ASTM D4253 and D4254).

given in Table 5-1 are for typical pipe zone embedment materials. For pipe zone embedment materials with a finer grain size, use the D_f value of sand with moderate to high compaction.

5.7.2.2 Long-term, ring-bending strain, S_b. The long-term, ring-bending strain varies for different products, depending on materials and type of construction used in the pipe wall. Long-term, ring-bending strain should be determined as defined in ANSI/AWWA Standard C950.

5.7.2.3 Bending design factor. Prudent design of pipe to withstand bending requires consideration of two separate design factors.

The first design consideration is comparison of initial deflection at failure to the maximum allowed installed deflection. The ring stiffness test (level B) in ANSI/AWWA Standard C950 subjects a pipe ring to deflections far exceeding those permitted in use. This test requirement demonstrates a design factor of at least 2.5 on initial bending strain.

The second design factor is the ratio of long-term bending stress or strain to the bending stress or strain at the maximum allowable long-term deflection. For fiberglass pipe design, this minimum design factor is 1.5.

5.7.3 Deflection

Buried pipe should be installed in a manner that will ensure that external loads will not cause a long-term decrease in the vertical diameter of the pipe exceeding the maximum allowable deflection ($\Delta y_a/D$) established in Sec. 5.7.2 or the permitted deflection ($\delta d/D$), as required by the engineer or manufacturer, whichever is less. This requirement may be stated as follows:

$$\Delta y/D \leq \delta d/D \leq \Delta y_a/D \qquad \qquad \Delta y/D \leq \delta d/D \leq \Delta y_a/D \qquad (5\text{-}7)$$

Where:
$\Delta y/D$ = predicted vertical pipe deflection, fraction of mean diameter

Where:
$\Delta y/D$ = predicted vertical pipe deflection, fraction of mean diameter

52 FIBERGLASS PIPE DESIGN

$\delta d/D$ = permitted vertical pipe deflection, fraction of mean diameter

$\Delta y_a/D$ = maximum allowable vertical pipe deflection, fraction of mean diameter

$\delta d/D$ = permitted vertical pipe deflection, fraction of mean diameter

$\Delta y_a/D$ = maximum allowable vertical pipe deflection, fraction of mean diameter

$$\frac{\Delta_y}{D} = \frac{(D_L W_c + W_L)K_x}{0.149\text{PS} + 0.061 M_s} \qquad \frac{\Delta_y}{D} = \frac{(D_L W_c + W_L)K_x}{149\text{PS} + 61{,}000 M_s} \qquad (5\text{-}8)$$

Where:

D_L = deflection lag factor to compensate for the time-consolidation rate of the soil, dimensionless

W_c = vertical soil load on pipe, psi

W_L = live load on pipe, psi

K_x = bedding coefficient, dimensionless

PS = pipe stiffness, psi

M_s = composite soil constrained modulus, psi

Where:

D_L = deflection lag factor to compensate for the time-consolidation rate of the soil, dimensionless

W_c = vertical soil load on pipe, N/m^2

W_L = live load on pipe, N/m^2

K_x = bedding coefficient, dimensionless

PS = pipe stiffness, kPa

M_s = composite soil constrained modulus, MPa

5.7.3.1 Deflection calculations. Design calculations that require deflection as an input parameter should show the predicted deflection $\Delta y/D$ as well as the maximum allowable deflection $\Delta y_a/D$ at which the allowable design stress or strain is not exceeded. The maximum permitted deflection $\delta d/D$ should be used in all design calculations.

5.7.3.2 Deflection prediction. When installed in the ground, all flexible pipe will undergo deflection, defined here to mean a decrease in vertical diameter. The amount of deflection is a function of the soil load, live load, native soil characteristics at pipe elevation, pipe embedment material and density, trench width, haunching, and pipe stiffness. Many theories have been proposed to predict deflection levels; however, in actual field conditions, pipe deflections may vary from calculated values because the actual installation achieved may vary from the installation planned. These variations include the inherent variability of native ground conditions and variations in methods, materials, and equipment used to install a buried pipe.

Field personnel responsible for pipe installation must follow procedures designed to ensure that the long-term pipe deflection is less than Δy_a as determined in Sec. 5.7.2, or as required by the engineer or manufacturer, whichever is less. As presented previously and as augmented by information provided in the following sections, Eq 5-8 serves as a guideline for estimating the expected level of short-term and long-term deflection that can be anticipated in the field. Equation 5-8 is a form of the Iowa formula, first published by Spangler[*] in 1941. This equation is the best known and documented of a multitude of deflection-prediction equations that have been proposed. As presented in this chapter, the Iowa formula treats the major aspects of pipe–soil interaction with sufficient accuracy to produce reasonable estimates of load-induced field deflection levels.

---

[*] Spangler, M.G., and R.L. Handy. *Soil Engineering.* New York: Harper & Row (4th ed., 1982).

Pipe deflection due to self-weight and initial ovalization due to pipe backfill embedment placement and compaction are not addressed by this method. These deflections are typically small for pipe stiffnesses above 9 psi to 18 psi (62 kPa to 124 kPa) (depending on installation conditions). For pipe stiffnesses below these values, consideration of these items may be required to achieve an accurate deflection prediction.

Application of this method is based on the assumption that the design values used for bedding, backfill, and compaction levels will be achieved with good practice and with appropriate equipment in the field. Experience has shown that deflection levels of any flexible conduit can be higher or lower than predicted by calculation if the design assumptions are not achieved.

5.7.3.3 Deflection lag factor, D_L. The deflection lag factor converts the immediate deflection of the pipe to the deflection of the pipe after many years. The primary cause of increasing pipe deflection with time is the increase in overburden load as soil "arching" is gradually lost. The vast majority of this phenomenon occurs during the first few weeks or months of burial and may continue for some years, depending on the frequency of wetting and drying cycles, surface loads, and the amount of original compaction of the final backfill. Secondary causes of increasing pipe deflection over time are the time-related consolidation of the pipe zone embedment and the creep of the native soil at the sides of the pipe. These causes are generally of much less significance than increasing load and may not contribute to the deflection for pipes buried in relatively stiff native soils with dense granular pipe zone surrounds. For long-term deflection prediction, a D_L value greater than 1.00 is appropriate.

5.7.3.4 Bedding coefficient, K_x. The bedding coefficient reflects the degree of support provided by the soil at the bottom of the pipe and over which the bottom reaction is distributed. Assuming an inconsistent haunch achievement (typical direct bury condition), a K_x value of 0.1 should be used. For uniform-shaped bottom support, a K_x value of 0.083 is appropriate.

5.7.3.5 Vertical soil load on the pipe, W_c. The long-term vertical soil load on the pipe may be considered as the weight of the rectangular prism of soil directly above the pipe. The soil prism would have a height equal to the depth of earth cover and a width equal to the pipe outside diameter.

$$W_c = \frac{\gamma_s H}{144} \qquad\qquad W_c = \gamma_s H \qquad (5\text{-}9)$$

Where:

W_c = vertical soil load, psi

γ_s = unit weight of overburden, lb/ft^3

H = burial depth to top of pipe, ft

Where:

W_c = vertical soil load, N/m^2

γ_s = unit weight of overburden, N/m^3

H = burial depth to top of pipe, m

5.7.3.6 Live loads on the pipe, W_L. The following calculations may be used to compute the live load on the pipe for surface traffic (see Figure 5-1). The procedure is based on the requirements of the *AASHTO LRFD* (American Association of State Highway and Transportation Officials load-and-resistance factor design philosophy) *Bridge Design Specification*, second edition, 1999. These calculations consider a single-axle truck traveling perpendicular to the pipe on an unpaved surface or a road with flexible pavement. With the inclusion of the multiple presence factor (M_p), the

54   FIBERGLASS PIPE DESIGN

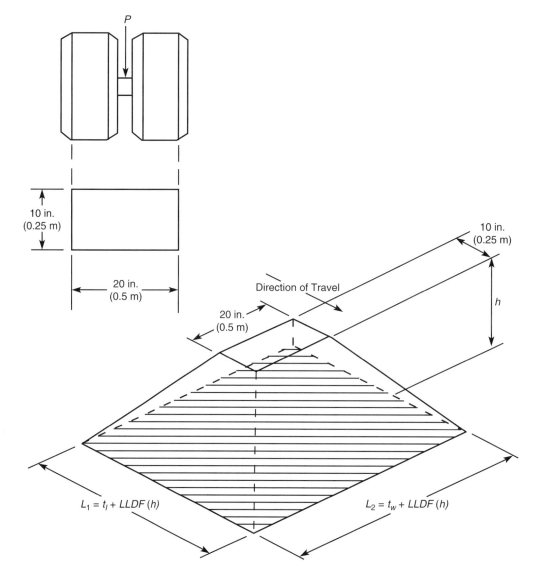

NOTE: For $h > 45$ in. (1.14 m), see part 2 of the L_2 formula. (Change accounts for overlapping influence areas from adjacent wheel loads.)

Figure 5-1   Distribution of AASHTO HS-20 or HS-25 live load through granular fill for $h \leq 45$ in. (1.14 m)

previous conditions generally control and may be assumed to yield acceptably conservative load estimates.

$$W_L = \frac{M_p P I_f}{(L_1)(L_2)} \qquad\qquad W_L = \frac{M_p P I_f}{(L_1)(L_2)} \qquad (5\text{-}10)$$

Where:
W_L = live load on pipe, psi

Where:
W_L = live load on pipe, N/m^2

M_P = multiple presence factor = 1.2

P = wheel load magnitude

 = 16,000 lb for AASHTO HS-20 truck

 = 20,000 lb for AASHTO HS-25 truck

I_f = impact factor

L_1 = load width parallel to direction of travel, in.

L_2 = load width perpendicular to direction of travel, in.

M_P = multiple presence factor = 1.2

P = wheel load magnitude

 = 71,300 N for AASHTO HS-20 truck

 = 89,000 N for AASHTO HS-25 truck

I_f = impact factor

L_1 = load width parallel to direction of travel, m

L_2 = load width perpendicular to direction of travel, m

$$I_f = 1 + 0.33\,[(96 - h)/96] \geq 1.0 \qquad\qquad I_f = 1 + 0.33\,[(2.44 - h)/2.44] \geq 1.0 \quad (5\text{-}11)$$

Where:

h = depth of cover, in.

Where:

h = depth of cover, m

$$L_1 = t_l + LLDF\,(h) \qquad\qquad L_1 = t_l + LLDF\,(h) \quad (5\text{-}12)$$

Where:

t_l = length of tire footprint = 10 in.

$LLDF$ = factor to account for live load distribution with depth of fill

 = 1.15 for backfills SC1 and SC2

 = 1.0 for all other backfills

Where:

t_l = length of tire footprint = 0.25 m

$LLDF$ = factor to account for live load distribution with depth of fill

 = 1.15 for backfills SC1 and SC2

 = 1.0 for all other backfills

$$\text{If } h \leq h_{int} \qquad\qquad \text{If } h \leq h_{int} \quad (5\text{-}13)$$
$$L_2 = t_w + LLDF\,(h) \qquad\qquad L_2 = t_w + LLDF\,(h)$$

Where:

t_w = width of tire footprint = 20 in.

Where:

t_w = width of tire footprint = 0.5 m

$$\text{If } h > h_{int} \qquad\qquad \text{If } h > h_{int} \quad (5\text{-}14)$$
$$L_2 = [t_w + 72\text{ in.} + LLDF\,(h)]/2 \qquad\qquad L_2 = [t_w + 1.83\text{ m} + LLDF\,(h)]/2$$

Where:

h_{int} = depth at which load from wheels interacts

Where:

h_{int} = depth at which load from wheels interacts

$$h_{int} = (72\text{ in.} - t_w)/LLDF \qquad\qquad h_{int} = (1.83\text{ m} - t_w)/LLDF \quad (5\text{-}15)$$

Calculation notes:
1. Equations as shown are for h in inches (meters).
2. AASHTO also specifies a surface lane load of 640 lb/ft (9,350 N/m) over a 10-ft (3-m) lane width. This lane load is ignored in these calculations because it has only a small effect on the total live load and may be added by the engineer if deemed appropriate.
3. The above calculation method assumes that the live load extends over the full diameter of the pipe. This may be conservative for large-diameter pipe under low fills. To account for this, the calculated live load pressure on the pipe may be reduced by the ratio L_1/OD if the truck is moving across the pipe and L_1 < OD or by L_2/OD if the truck is moving parallel to the pipe and L_2 < OD. The OD is the outside diameter of the pipe in inches (millimeters).
4. For depths of fill less than 2 ft (0.6 m) or for live load magnitudes greater than HS-25, it may be necessary to consider the local live load effects at the crown of the pipe. Such an analysis is beyond the scope of this manual.
5. The previous calculation is for single-axle trucks. Design for tandem-axle trucks may use the same procedures; however, the following substitutions for L_1 should be used if both axles load the pipe at the same time.

$$L_1 = [\text{axle spacing} + t_w + LLDF\,(h)]/2 \qquad L_1 = [\text{axle spacing} + t_w + LLDF\,(h)]/2 \quad (5\text{-}16)$$

Tandem-axle wheel loads are usually lighter than HS-20 or HS-25 trucks; for example, the AASHTO LRFD design tandem has a 12,500-lb (55,700-N) wheel load.

6. Rigid pavements dramatically reduce live load effects on concrete pipe. The Portland Cement Association developed a calculation method to consider loads transmitted through concrete pavements (*Vertical Pressure on Concrete Culverts Under Wheel Loads on Concrete Pavement Slabs*, Portland Cement Association, Publication ST-65, 1951) that is still in use today and is suitable for computing live loads on fiberglass pipe under rigid pavements. The same method is also presented in the *Concrete Pipe Handbook*, published by the American Concrete Pipe Association.

Table 5-2 presents computed live loads for AASHTO HS-20 and HS-25 single-axle trucks based on an $LLDF$ of 1.15 (for granular fills). The loads shown assume that the load extends over the full diameter of the pipe. This assumption will not be true for large-diameter pipes with shallow covers. Loads for this condition may be lower. See calculation note 3 for guidance on appropriate adjustments.

Also included in Table 5-2 are live loads from Cooper E80 railroad loading (taken from the *Concrete Pipe Design Manual*).

Figures 5-2 through 5-4 graphically depict the AASHTO truck and Cooper E80 railroad live loads shown in Table 5-2.

5.7.3.7 Pipe stiffness, PS. The pipe stiffness can be determined by conducting parallel-plate loading tests in accordance with ASTM D2412. During the parallel-plate loading test, deflection due to loads on the top and bottom of the pipe is measured, and pipe stiffness is calculated from the following equation:

$$PS = F/\Delta y_t \qquad\qquad PS = 1{,}000\,F/\Delta y_t \quad (5\text{-}17)$$

BURIED PIPE DESIGN 57

Table 5-2 AASHTO HS-20, HS-25, and Cooper E80 live loads (psi)

AASHTO Live Loads		HS-20		HS-25		Cooper E80 Live Loads			
Depth		W_L		W_L		Depth		W_L	
ft	m	psi	kPa	psi	kPa	ft	m	psi	kPa
2	0.6	13.4	92	16.8	116	3	0.9	16	110
2.5	0.8	9.7	67	12.2	84	4	1.2	14.1	97
3	0.9	7.4	51	9.2	63	5	1.5	12.2	84
4	1.2	4.7	32	5.9	41	6	1.8	10.5	72
5	1.5	3.4	23	4.2	29	7	2.1	9	62
6	1.8	2.6	18	3.2	22	8	2.4	7.7	53
8	2.4	1.6	11	2	14	10	3.0	5.7	39
10	3.0	1.1	7.6	1.4	10	12	3.7	4.6	32
12	3.7	0.8	5.5	1.1	7.6	15	4.6	3.4	23
15	4.6	0.6	4.1	0.7	4.8	20	6.1	2.2	15
20	6.1	0.4	2.8	0.5	3.4	25	7.6	1.5	10
28	8.5	0.2	1.4	0.25	1.8	30	9.1	1.1	7.6
40	12.2	0.1	0.7	0.1	0.7	40	12.2	0.6	4.1

NOTE: Cooper E80 as defined by the American Railway Engineers and Maintenance-of-Way Association (AREMA) in their *Manual of Railway Engineering*.

Where:

PS = pipe stiffness, psi
F = load per unit length, lb/in.
Δy_t = vertical pipe deflection, in., when tested by ASTM D2412 with a vertical diameter reduction of 5%

Where:

PS = pipe stiffness, kPa
F = load per unit length, N/mm
Δy_t = vertical pipe deflection, mm, when tested by ASTM D2412 with a vertical diameter reduction of 5%

Pipe stiffness may also be determined by the pipe dimensions and material properties using Eq 5-18:

$$PS = \frac{EI}{0.149(r + \Delta y_t/2)^3} \qquad PS = \frac{EI \times 10^6}{0.149(r + \Delta y_t/2)^3} \qquad (5\text{-}18)$$

Where:

E = ring flexural modulus, psi
I = moment of inertia of unit length, in.^{4}/in.
  = $(t_t)^3/12$
r = mean pipe radius, in.
  = (OD – t)/2

Where:

E = ring flexural modulus, GPa
I = moment of inertia of unit length, mm^4/mm
  = $(t_t)^3/12$
r = mean pipe radius, mm
  = (OD – t)/2

58   FIBERGLASS PIPE DESIGN

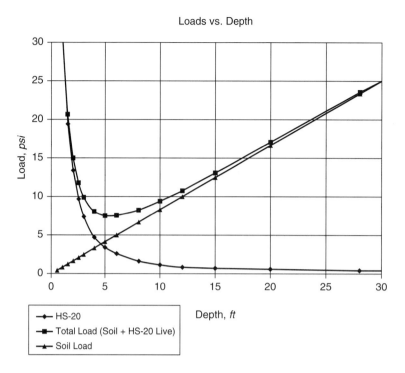

Figure 5-2   AASHTO HS-20 live load, soil load (120 pcf), and total load graph

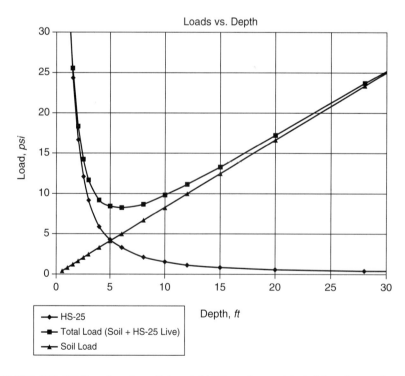

Figure 5-3   AASHTO HS-25 live load, soil load (120 pcf), and total load graph

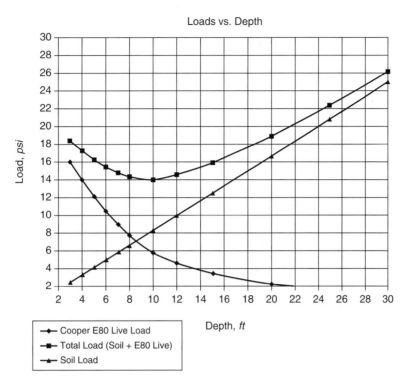

Figure 5-4 Cooper E80 live load, soil load (120 pcf), and total load graph

5.7.3.8 Constrained soil modulus, M_s. The vertical loads on a flexible pipe cause a decrease in the vertical diameter and an increase in the horizontal diameter. The horizontal movement develops a passive soil resistance that helps support the pipe. The passive soil resistance varies depending on the soil type and the degree of compaction of the pipe zone backfill material, native soil characteristics at pipe elevation, cover depth, and trench width (see Table 5-3).

The historical parameter used to characterize the soil stiffness in design of flexible pipe is the modulus of soil reaction E'. In 2000, AASHTO adopted new soil stiffness values for backfill materials around thermoplastic pipe, including changing the soil design parameter from E' to the constrained modulus M_s. This change is based on the work of McGrath (1998). Design values of the constrained modulus are presented in Table 5-4. The table shows that M_s increases with depth of fill, which reflects the increased confining pressure. This is a well-known soil behavior. At moderate depths of fill, the values of M_s are close to the E' values proposed by Howard (1977, 1996). In design for deflection control, M_s may be substituted directly for E' in the Iowa formula.

To determine M_s for a buried pipe, separate M_s values for the native soil M_{sn} and the pipe backfill surround M_{sb} must be determined and then combined using Eq 5-19. Special cases are discussed later in this chapter.

$$M_s = S_c M_{sb} \qquad\qquad M_s = S_c M_{sb} \qquad (5\text{-}19)$$

Where:

M_s = composite constrained soil modulus, psi

Where:

M_s = composite constrained soil modulus, MPa

60   FIBERGLASS PIPE DESIGN

S_c = soil support combining factor from Table 5-5, dimensionless

M_{sb} = constrained soil modulus of the pipe zone embedment from Table 5-4, psi

To use Table 5-5 for S_c, the following values are required:

M_{sn} = constrained soil modulus of the native soil at pipe elevation, psi (see Table 5-6)

B_d = trench width at pipe springline, in.

S_c = soil support combining factor from Table 5-5, dimensionless

M_{sb} = constrained soil modulus of the pipe zone embedment from Table 5-4, MPa

To use Table 5-5 for S_c, the following values are required:

M_{sn} = constrained soil modulus of the native soil at pipe elevation, MPa (see Table 5-6)

B_d = trench width at pipe springline, mm

5.7.4 Combined Loading

The maximum stress or strain resulting from the combined effects of internal pressure and deflection should meet Eq 5-20 and Eq 5-21 or Eq 5-22 and Eq 5-23 as follows:

For stress basis HDB:

$$\frac{\sigma_{pr}}{\text{HDB}} \leq \frac{1 - \left(\frac{\sigma_b r_c}{S_b E}\right)}{FS_{pr}} \qquad \frac{\sigma_{pr}}{\text{HDB}} \leq \frac{1 - \left(\frac{\sigma_b r_c}{S_b E \times 10^3}\right)}{FS_{pr}} \qquad (5\text{-}20)$$

$$\frac{\sigma_b r_c}{S_b E} \leq \frac{1 - \left(\frac{\sigma_{pr}}{\text{HDB}}\right)}{FS_b} \qquad \frac{\sigma_b r_c}{S_b E \times 10^3} \leq \frac{1 - \left(\frac{\sigma_{pr}}{\text{HDB}}\right)}{FS_b} \qquad (5\text{-}21)$$

For strain basis HDB:

$$\frac{\varepsilon_{pr}}{\text{HDB}} \leq \frac{1 - \left(\frac{\varepsilon_b r_c}{S_b}\right)}{FS_{pr}} \qquad \frac{\varepsilon_{pr}}{\text{HDB}} \leq \frac{1 - \left(\frac{\varepsilon_b r_c}{S_b}\right)}{FS_{pr}} \qquad (5\text{-}22)$$

$$\frac{\varepsilon_b r_c}{S_b} \leq \frac{1 - \left(\frac{\varepsilon_{pr}}{\text{HDB}}\right)}{FS_b} \qquad \frac{\varepsilon_b r_c}{S_b} \leq \frac{1 - \left(\frac{\varepsilon_{pr}}{\text{HDB}}\right)}{FS_b} \qquad (5\text{-}23)$$

Where:

FS_{pr} = pressure design factor, 1.8

FS_b = bending design factor, 1.5

Where:

FS_{pr} = pressure design factor, 1.8

FS_b = bending design factor, 1.5

Table 5-3 Soil classification chart

Criteria for Assigning Group Symbols and Group Names Using Laboratory Tests[a]				Soil Classification	
				Group Symbol	Group Name[b]
Coarse-grained soils More than 50% retained on No. 200 sieve	Gravels More than 50% of coarse fraction retained on No. 4 sieve	Clean gravels Less than 5% fines[c]	$Cu \geq 4$ and $1 \leq Cc \leq 3$[e]	GW	Well-graded gravel[f]
			$Cu < 4$ and/or $1 > Cc > 3$[e]	GP	Poorly graded gravel[f]
		Gravels with fines More than 12% fines[c]	Fines classify as ML or MH	GM	Silty gravel[f,g,h]
			Fines classify as CL or CH	GC	Clayey gravel[f,g,h]
	Sands 50% or more of coarse fraction passes No. 4 sieve	Clean sands Less than 5% fines[d]	$Cu \geq 6$ and $1 \leq Cc \leq 3$[e]	SW	Well-graded sand[i]
			$Cu < 6$ and/or $1 > Cc > 3$[e]	SP	Poorly graded sand[i]
		Sands with fines More than 12% fines[d]	Fines classify as ML or MH	SM	Silty sand[g,h,i]
			Fines classify as CL or CH	SC	Clayey sand[g,h,i]
Fine-grained soils 50% or more passes the No. 200 sieve	Silts and clays Liquid limit less than 50	Inorganic	PI > 7 and plots on or above "A" line[j]	CL	Lean clay[k,l,m]
			PI < 4 or plots below "A" line[j]	ML	Silt[k,l,m]
		Organic	Liquid limit—oven dried / Liquid limit—not dried < 0.75	OL	Organic clay[k,l,m,n] / Organic silt[k,l,m,o]
	Silts and clays Liquid limit 50 or more	Inorganic	PI plots on or above "A" line	CH	Fat clay[k,l,m]
			PI plots below "A" line	MH	Elastic silt[k,l,m]
		Organic	Liquid limit—oven dried / Liquid limit—not dried < 0.75	OH	Organic clay[k,l,m,p] / Organic silt[k,l,m,q]
Highly organic soils		Primarily organic matter, dark in color, and organic odor		PT	Peat

[a] Based on the material passing the 3-in. (75-mm) sieve.

[b] If field sample contained cobbles and/or boulders, add "with cobbles and/or boulders" to group name.

[c] Gravels with 5% to 12% fines require dual symbols:
GW–GM well-graded gravel with silt
GW–GC well-graded gravel with clay
GP–GM poorly graded gravel with silt
GP–GC poorly graded gravel with clay

[d] Sands with 5% to 12% fines require dual symbols:
SW–SM well-graded sand with silt
SW–SC well-graded sand with clay
SP–SM poorly graded sand with silt
SP–SC poorly graded sand with clay

[e] $Cu = D_{60}/D_{10}$

$$Cc = \frac{(D_{30})^2}{D_{10} \times D_{60}}$$

[f] If soil contains ≥ 15% sand, add "with sand" to group name.

[g] If fines classify as CL–ML, use dual symbol GC–GM or SC–SM.

[h] If fines are organic, add "with organic fines" to group name.

[i] If soil contains ≥ 15% gravel, add "with gravel" to group name.

[j] If the Atterberg limits (liquid limit and plasticity index) plot in hatched area on plasticity chart, soil is a CL–ML, silty clay.

[k] If soil contains 15% to 29% plus No. 200, add "with sand" or "with gravel," whichever is predominant.

[l] If soil contains ≥ 30% plus No. 200, predominantly sand, add "sandy" to group name.

[m] If soil contains ≥ 30% plus No. 200, predominantly gravel, add "gravelly" to group name.

[n] PI ≥ 4 and plots on or above "A" line.

[o] PI ≤ 4 or plots below "A" line.

[p] PI plots on or above "A" line.

[q] PI plots below "A" line

Source: ASTM D2487. Reprinted with permission from the Annual Book of ASTM Standards. *Copyright ASTM, 100 Barr Harbor Dr., West Conshohocken, PA 19428-2959.*

NOTE: ASTM D2487 allows the use of "borderline" symbols when test results indicate that the soil classification is close to another group. The borderline condition is indicated by an en dash between the two symbols, for example, CL–CH.

Table 5-4 M_{sb} based on soil type and compaction condition

Inch-Pound Units

Vertical Stress Level (see note 5), psi	Depth for γ_s = 120 pcf, ft	Stiffness Categories 1 and 2 (SC1, SC2)			
		SPD100, psi	SPD95, psi	SPD90, psi	SPD85, psi
1	1.2	2,350	2,000	1,275	470
5	6	3,450	2,600	1,500	520
10	12	4,200	3,000	1,625	570
20	24	5,500	3,450	1,800	650
40	48	7,500	4,250	2,100	825
60	72	9,300	5,000	2,500	1,000
		Stiffness Category 3 (SC3)			
1	1.2		1,415	670	360
5	6		1,670	740	390
10	12		1,770	750	400
20	24		1,880	790	430
40	48		2,090	900	510
60	72		2,300	1,025	600
		Stiffness Category 4 (SC4)			
1	1.2		530	255	130
5	6		625	320	175
10	12		690	355	200
20	24		740	395	230
40	48		815	460	285
60	72		895	525	345

Metric

Vertical Stress Level (see note 5), kPa	Depth for Soil Density = 18.8 kN/m^3, m	Stiffness Categories 1 and 2 (SC1, SC2)			
		SPD100, MPa	SPD95, MPa	SPD90, MPa	SPD85, MPa
6.9	0.4	16.2	13.8	8.8	3.2
34.5	1.8	23.8	17.9	10.3	3.6
69	3.7	29	20.7	11.2	3.9
138	7.3	37.9	23.8	12.4	4.5
276	14.6	51.7	29.3	14.5	5.7
414	22	64.1	34.5	17.2	6.9

Table continued on next page

Table 5-4 M_{sb} based on soil type and compaction condition, *continued*

Metric, *continued*

Vertical Stress Level (see note 5), kPa	Depth for Soil Density = 18.8 kN/m³, m	Stiffness Category 3 (SC3)			
		SPD100, MPa	SPD95, MPa	SPD90, MPa	SPD85, MPa
6.9	0.4		9.8	4.6	2.5
34.5	1.8		11.5	5.1	2.7
69	3.7		12.2	5.2	2.8
138	7.3		13	5.4	3
276	14.6		14.4	6.2	3.5
414	22		15.9	7.1	4.1
		Stiffness Category 4 (SC4)			
6.9	0.4		3.7	1.8	0.9
34.5	1.8		4.3	2.2	1.2
69	3.7		4.8	2.5	1.4
138	7.3		5.1	2.7	1.6
276	14.6		5.6	3.2	2
414	22		6.2	3.6	2.4

NOTES:

1. SC1 soils have the highest stiffness and require the least amount of compactive energy to achieve a given density. SC5 soils, which are not recommended for use as backfill, have the lowest stiffness and require substantial effort to achieve a given density. Soil stiffness categories are explained in chapter 6.

2. SC1 soils have higher stiffness than SC2 soils, but data on specific soil stiffness values is not available at the current time. Until such data is available, the soil stiffness of dumped SC1 soils can be taken equivalent to SC2 soils compacted to 90% of maximum standard Proctor density (SPD90), and the soil stiffness of compacted SC1 soils can be taken equivalent to SC2 soils compacted to 100% of maximum standard Proctor density (SPD100). Even if dumped, SC1 materials should always be worked into the haunch zone, see Sec. 6.7.3.

3. The soil types SC1 to SC5 are defined in Table 6-1. Specific soil groups that fall into these categories, based on ASTM D2487 and AASHTO M145, are also listed in Table 6-1.

4. The numerical suffix to the SPD (standard Proctor density) indicates the compaction level of the soil as a percentage of maximum dry density determined in accordance with ASTM D698 or AASHTO T-99.

5. Vertical stress level is the vertical effective soil stress at the springline elevation of the pipe. It is normally computed as the design soil unit weight times the depth of fill. Buoyant unit weight should be used below the groundwater level.

6. Engineers may interpolate intermediate values of M_{sb} for vertical stress levels not shown on the table.

7. For pipe installed below the water table, the modulus should be corrected for reduced vertical stress due to buoyancy and by an additional factor of 1.00 for SC1 and SC2 soils with SPD of ≥95, 0.85 for SC2 soils with SPD of 90, 0.70 for SC2 soils with SPD of 85, 0.50 for SC3 soils, and 0.30 for SC4 soils.

8. It is recommended to embed pipe with stiffness of 9 psi (62 kPa) or less only in SC1 or SC2 soils.

Table 5-5 Values for the soil support combining factor S_c

M_{sn}/M_{sb}	$B_d/D = 1.25$	$B_d/D = 1.5$	$B_d/D = 1.75$	$B_d/D = 2$	$B_d/D = 2.5$	$B_d/D = 3$	$B_d/D = 4$	$B_d/D = 5$
0.005	0.02	0.05	0.08	0.12	0.23	0.43	0.72	1.00
0.01	0.03	0.07	0.11	0.15	0.27	0.47	0.74	1.00
0.02	0.05	0.10	0.15	0.20	0.32	0.52	0.77	1.00
0.05	0.10	0.15	0.20	0.27	0.38	0.58	0.80	1.00
0.1	0.15	0.20	0.27	0.35	0.46	0.65	0.84	1.00
0.2	0.25	0.30	0.38	0.47	0.58	0.75	0.88	1.00
0.4	0.45	0.50	0.56	0.64	0.75	0.85	0.93	1.00
0.6	0.65	0.70	0.75	0.81	0.87	0.94	0.98	1.00
0.8	0.84	0.87	0.90	0.93	0.96	0.98	1.00	1.00
1	1.00	1.00	1.00	1.00	1.00	1.00	1.00	1.00
1.5	1.40	1.30	1.20	1.12	1.06	1.03	1.00	1.00
2	1.70	1.50	1.40	1.30	1.20	1.10	1.05	1.00
3	2.20	1.80	1.65	1.50	1.35	1.20	1.10	1.00
≥5	3.00	2.20	1.90	1.70	1.50	1.30	1.15	1.00

NOTE: In-between values of S_c may be determined by straight-line interpolation from adjacent values.

Table 5-6 Values for the constrained modulus of the native soil at pipe zone elevation

Native In Situ Soils*						
Granular		Cohesive			M_{sn}	
		q_u				
Blows/ft[†] (0.3 m)	Description	tons/sf	kPa	Description	psi	MPa
>0–1	very, very loose	>0–0.125	0–13	very, very soft	50	0.34
1–2	very loose	0.125–0.25	13–25	very soft	200	1.4
2–4		0.25–0.50	25–50	soft	700	4.8
4–8	loose	0.50–1.0	50–100	medium	1,500	10.3
8–15	slightly compact	1.0–2.0	100–200	stiff	3,000	20.7
15–30	compact	2.0–4.0	200–400	very stiff	5,000	34.5
30–50	dense	4.0–6.0	400–600	hard	10,000	69.0
>50	very dense	>6.0	>600	very hard	20,000	138.0

* The constrained modulus M_{sn} for rock is ≥ 50,000 psi (345 MPa).

† Standard penetration test per ASTM D1586.

NOTES: For embankment installation $M_{sb} = M_{sn} = M_s$.

M_s special cases:

Geotextiles—When a geotextile pipe zone wrap is used, M_{sn} values for poor soils can be greater than those shown in this table.

Solid sheeting—When permanent solid sheeting designed to last the life of the pipeline is used in the pipe zone, M_s shall be based solely on M_{sb}.

Cement-stabilized sand—When cement-stabilized sand is used as the pipe zone surround, initial deflections shall be based on a sand installation and the long-term M_{sb} = 25,000 psi (172 MPa) (Typical mix ratio is 1 sack of cement per ton or 1.5 sacks of cement per cubic yard of mix.).

σ_{pr} = working stress due to internal pressure, psi

$$= \frac{P_w D}{2t}$$

σ_b = bending stress due to the maximum permitted deflection, psi

$$= D_f E \left(\frac{\delta d}{D}\right)\left(\frac{t_t}{D}\right)$$

r_c = rerounding coefficient, dimensionless

$= 1 - P_w/435$ (where $P_w \le$ 435 psi)

ε_{pr} = working strain due to internal pressure, in./in.

$$= \frac{P_w D}{2t E_H}$$

ε_b = bending strain due to maximum permitted deflection, in./in.

$$= D_f \left(\frac{\delta d}{D}\right)\left(\frac{t_t}{D}\right)$$

δd = maximum permitted long-term installed deflection, in.

σ_{pr} = working stress due to internal pressure, MPa

$$= \frac{P_w D}{2t}$$

σ_b = bending stress due to the maximum permitted deflection, MPa

$$= D_f E \left(\frac{\delta d}{D}\right)\left(\frac{t_t}{D}\right)$$

r_c = rerounding coefficient, dimensionless

$= 1 - P_w/3{,}000$ (where $P_w \le$ 3,000 kPa)

ε_{pr} = working strain due to internal pressure, mm/mm

$$= \frac{P_w D}{2t E_H}$$

ε_b = bending strain due to maximum permitted deflection, mm/mm

$$= D_f \left(\frac{\delta d}{D}\right)\left(\frac{t_t}{D}\right)$$

δd = maximum permitted long-term installed deflection, mm

5.7.5 Buckling

5.7.5.1 Buckling theory. Buried pipe is subjected to radial external loads composed of vertical loads and the hydrostatic pressure of groundwater and internal vacuum, if the latter two are present. External radial pressure sufficient to buckle buried pipe is many times higher than the pressure causing buckling of the same pipe in a fluid environment, due to the restraining influence of the soil.

5.7.5.2 Buckling calculations. The summation of appropriate external loads should be equal to or less than the allowable buckling pressure. The allowable buckling pressure q_a is determined by the following equation:

$$q_a = \frac{(1.2 C_n)(EI)^{0.33}(\varphi_s M_s k_\upsilon)^{0.67} R_h}{(FS) r} \qquad q_a = \frac{(1.2 C_n)(EI)^{0.33}(\varphi_s 10^6 M_s k_\upsilon)^{0.67} R_h}{(FS) r} \quad (5\text{-}24\mathrm{a})$$

Where:

q_a = allowable buckling pressure, psi

FS = design factor, 2.5

C_n = scalar calibration factor to account for some nonlinear effects = 0.55

Where:

q_a = allowable buckling pressure, kPa

FS = design factor, 2.5

C_n = scalar calibration factor to account for some nonlinear effects = 0.55

φ_s = factor to account for variability in stiffness of compacted soil; suggested value is 0.9

k_υ = modulus correction factor for Poisson's ratio, υ, of the soil
   = $(1 + \upsilon)(1 - 2\upsilon)/(1 - \upsilon)$; in the absence of specific information, it is common to assume $\upsilon = 0.3$ giving $k_\upsilon = 0.74$

R_h = correction factor for depth of fill
   = $11.4/(11 + D/h)$ [= $11.4/(11 + D/1{,}000\,h)$]

h = height of ground surface above top of pipe, in. [m]

An alternate form of Eq 5-24a is:

$$q_a = \left(\frac{1}{FS}\right)[1.2 C_n (0.149 \text{PS})^{0.33}](\varphi_s M_s k_\upsilon)^{0.67} R_h$$

$$q_a = \left(\frac{1}{FS}\right)[1.2 C_n (0.149 \text{PS})^{0.33}](\varphi_s 10^6 M_s k_\upsilon)^{0.67} R_h \quad (5\text{-}24b)$$

Typical pipe installations. Satisfaction of the buckling requirement is assured for typical pipe installations by using the following equation:

$$\gamma_w h_w + R_w(W_c) + P_v \leq q_a \quad\quad [\gamma_w h_w + R_w(W_c)] \times 10^{-3} + P_v \leq q_a \quad (5\text{-}25)$$

Where:

γ_w = specific weight of water
   = 0.0361 lb/in.^{3} [9,800 N/m^3]

P_v = internal vacuum pressure (i.e., the atmospheric pressure less absolute pressure inside pipe), psi [kPa]

R_w = water buoyancy factor
   = $1 - 0.33(h_w/h)$ $[0 \leq h_w \leq h]$

h_w = height of water surface above the pipe top, in. [m]

If live loads are considered, satisfaction of the buckling requirement is ensured by:

$$\gamma_w h_w + R_w(W_c) + W_L \leq q_a \quad\quad [\gamma_w h_w + R_w(W_c) + W_L] \times 10^{-3} \leq q_a \quad (5\text{-}26)$$

Typically live load and internal vacuum are not considered simultaneously.

BURIED PIPE DESIGN 67

5.8 AXIAL LOADS

Factors that contribute to the development of axial stresses in buried pipe are (1) hoop expansion due to internal pressure, which causes axial tensile stresses whenever the pipe is axially restrained; (2) restrained thermal expansion and contraction; and (3) pipe "beam" bending that may be induced by uneven bedding, differential soil settlement, or subsidence of soil. The minimum requirements for axial strengths are as specified by Sec. 5.1.2.4 and Sec. 5.1.2.5 and Tables 11, 12, and 13 of ANSI/AWWA Standard C950. These requirements include service conditions in typical underground pipe installations with unrestrained joints that comply with the guidelines provided in chapter 6 of this manual and that have thrust blocks provided at bends, blanks, and valves in accordance with chapter 7 and pipe manufacturers' recommendations. When restrained joints are used, the pipe should be designed to accommodate the full magnitude of forces generated by internal pressure.

5.9 SPECIAL DESIGN CONSIDERATIONS

Pipe that meets the design requirements of ANSI/AWWA Standard C950 and Sec. 5.7 and that is installed in accordance with chapter 6 guidelines has adequate strength for service in usual buried applications. Special consideration should be made for the following conditions: (1) elevated temperature service; (2) broad temperature fluctuations; (3) shallow burial, where $H < 2$ ft (0.6 m); (4) uneven bedding or differential settlement of unstable native soils; (5) restrained tension joints; (6) extremely difficult construction conditions (for example, subaqueous installation); and (7) unusually high surface or construction loads.

5.10 DESIGN EXAMPLE

Example design calculations are presented in this section for a specific situation. For reference, the set of design conditions, pipe properties, and installation parameters assumed for this design example are presented in Table 5-7. This summary is not repeated in the body of the example design calculations.

The pipe material properties and characteristics presented in Table 5-7 have been assumed for illustrative purposes and should not be used as actual design values. Values for these parameters differ for various pipe constructions and materials and should be obtained from the manufacturer.

Step 1. Confirm pressure class (Eq 5-2).

$$P_c \leq \frac{(\text{HDB})(E_H)(2t)}{(FS)(D)} \text{ (strain basis)} \qquad P_c \leq \frac{(\text{HDB})(E_H)(2t)10^6}{(FS)(D)} \text{ (strain basis)}$$

$$P_c = 150 \text{ psi} \leq \qquad P_c = 1{,}000 \text{ kPa} \leq$$

$$\frac{(.0065)(1.8 \times 10^6)(2)(.61)}{1.8(36.69)} \qquad \frac{(.0065)(12.5)(2)(15.5)(10^6)}{1.8(908.5)}$$

$$\leq 216 \text{ psi} \therefore \text{OK} \qquad \leq 1{,}537 \text{ kPa} \therefore \text{OK}$$

Table 5-7 Conditions and parameters for design example

Parameter	Value, in.-lb	Units	Value, SI	Units
Design Conditions				
Nominal diameter	36	in.	900	mm
Working pressure	115	psi	800	kPa
Surge pressure	55	psi	375	kPa
Vacuum	14.7	psi	100	kPa
Cover depth				
Maximum	8	ft	2.5	m
Minimum	4	ft	1.2	m
Wheel load	20,000	lb	90,000	N
Service temperature	33 to 90	°F	1 to 32	°C
Pipe Characteristics				
Pressure class	150	psi	1,000	kPa
Nominal stiffness	36	psi	250	kPa
Inside diameter	36	in.	—	—
Outside diameter	—	—	924	mm
Wall thickness	0.65	in.	16.5	mm
Liner thickness	0.04	in.	1	mm
Reinforced wall thickness	0.61	in.	15.5	mm
Mean diameter	36.69	in.	908.5	mm
Hoop tensile modulus	1,800,000	psi	12.5	GPa
HDB	0.0065	in./in.	0.0065	mm/mm
S_b	0.012	in./in.	0.012	mm/mm
Maximum permitted deflection	5	%	5	%
Installation Parameters				
Native soil				
Description		Dense silty sand		
Soil modulus	10,000	psi	69	MPa
Backfill soil				
Description		moderately compacted sand (SC2 at 90 Proctor density)		
Soil modulus maximum depth	1,625	psi	11.2	MPa
Soil modulus minimum depth	1,500	psi	10.4	MPa
Specific weight	120	lb/ft^3	18,800	N/m^3
Groundwater effects				
Depth below grade	3	ft	1	m
Saturation reduction factor	0.75	—	0.75	—
Saturated soil modulus maximum depth	1,219	psi	8.4	MPa
Saturated soil modulus minimum depth	1,125	psi	7.8	MPa
Shape factor	5.5	—	5.5	—
Deflection lag factor	1.05	—	1.05	—
Deflection coefficient	0.1	—	0.1	—
Trench width	54	in.	1,350	mm

Step 2. Check working pressure (Eq 5-3).

$$P_c \geq P_w \qquad\qquad P_c \geq P_w$$

150 psi ≥ 115 psi ∴ OK $\qquad$ 1,000 kPa ≥ 800 kPa ∴ OK

Step 3. Check surge pressure (Eq 5-4).

$$P_c = \frac{(P_w + P_s)}{1.4} \qquad\qquad P_c = \frac{(P_w + P_s)}{1.4}$$

$$P_c = 150 \text{ psi} \geq \frac{(115 + 55)}{1.4} \qquad\qquad P_c = 1,000 \text{ kPa} \geq \frac{(800 + 375)}{1.4}$$

150 psi ≥ 122 psi ∴ OK $\qquad$ 1,000 kPa ≥ 840 kPa ∴ OK

Step 4. Calculate maximum allowable deflection (Eq 5-6).

$\varepsilon_b = (D_f)(\Delta y_a/D)(t_t/D) \leq (S_b)/FS$ $\qquad$ $\varepsilon_b = (D_f)(\Delta y_a/D)(t_t/D) \leq (S_b)/FS$

5.5(Δy_a/36.69)(.65/36.69) $\qquad\qquad$ 5.5(Δy_a/908.5)(16.5/908.5)
≤ (.012)/1.5 $\qquad\qquad\qquad\qquad\qquad$ ≤ (0.012)/1.5

0.0027 $\Delta y_a \leq 0.0080$ $\qquad\qquad\qquad$ 0.00011 $\Delta y_a \leq 0.0080$

maximum Δy_a = 2.96 in. $\qquad\qquad\qquad$ maximum Δy_a = 72.7 mm

maximum allowable deflection = (2.96/ $\qquad$ maximum allowable deflection = (72.7/
 36.69) × 100 $\qquad\qquad\qquad\qquad\qquad$ 908.5) × 100

= 8.1% $\qquad\qquad\qquad\qquad\qquad\qquad$ = 8.0%

permitted deflection = 5% ≤ 8.1% ∴ OK $\qquad$ permitted deflection = 5% ≤ 8.0% ∴ OK

Step 5. Calculate soils load (Eq 5-9).

$$W_c = \frac{\gamma_s H}{144} \qquad\qquad W_c = \gamma_s H$$

At minimum depth of 4 ft, $\qquad\qquad$ At minimum depth of 1.2 m,
$\quad W_c$ = 120(4)/144 $\qquad\qquad\qquad\qquad\quad W_c$ = 18,800(1.2)
$\qquad$ = 3.33 psi $\qquad\qquad\qquad\qquad\qquad\quad$ = 22,560 N/m^2

At maximum depth of 8 ft, $\qquad\qquad$ At maximum depth of 2.5 m,
$\quad W_c$ = 120(8)/144 $\qquad\qquad\qquad\qquad\quad W_c$ = 18,800(2.5)
$\qquad$ = 6.66 psi $\qquad\qquad\qquad\qquad\qquad\quad$ = 47,000 N/m^2

Step 6. Calculate live loads (Eq 5-10).

$$W_L = \frac{M_p P I_f}{(L_1)(L_2)} \qquad\qquad W_L = \frac{M_p P I_f}{(L_1)(L_2)}$$

70   FIBERGLASS PIPE DESIGN

Determine impact factor (Eq 5-11).

$I_f = 1 + 0.33[(96 - h)/96] \geq 1.0$ $I_f = 1 + 0.33[(2.44 - h)/2.44] \geq 1.0$

At minimum depth of 4 ft (48 in.),   At minimum depth of 1.2 m,
$I_f = 1 + 0.33[(96 - 48)/96] = 1.17$ $I_f = 1 + 0.33[(2.44 - 1.2)/2.44] \geq 1.17$
At maximum depth of 8 ft (96 in.),   At maximum depth of 2.5 m,
$I_f = 1 + 0.33[(96 - 96)/96] = 1.0$ $I_f = 1 + 0.33[(2.44 - 2.5)/2.44] \geq 1.0$

Determine L_1 (Eq 5-12).

$L_1 = t_l + LLDF\ (h)$ $L_1 = t_l + LLDF\ (h)$

At minimum depth of 4 ft (48 in.),   At minimum depth of 1.2 m,
$L_1 = 10 + 1.15\ (48) = 65$ in. $L_1 = .254 + 1.15\ (1.2) = 1.63$ m
At maximum depth of 8 ft (96 in.),   At maximum depth of 2.5 m,
$L_1 = 10 + 1.15\ (96) = 120$ in. $L_1 = .254 + 1.15\ (1.2) = 3.13$ m

Determine h_{int} (Eq 5-15).

$h_{int} = (72\ \text{in.} - t_w)/LLDF$ $h_{int} = (1.83\ \text{m} - t_w)/LLDF$

$h_{int} = (72\ \text{in.} - 20)/1.15 = 45.2$ in. $h_{int} = (1.83\ \text{m} - .508)/1.15 = 1.15$ m

At minimum depth of 4 ft (48 in.),   At minimum depth of 1.2 m, $h > h_{int}$
$h > h_{int}$ then calculate L_2 (Eq 5-14).   then calculate L_2 (Eq 5-14).
$L_2 = [t_w + 72\ \text{in.} + LLDF\ (h)]/2$ $L_2 = [t_w + 1.83\ \text{m} + LLDF\ (h)]/2$
$L_2 = [20 + 72 + 1.15\ (48)]/2 = 74$ in. $L_2 = [0.508 + 1.83 + 1.15(1.2)]/2 = 1.86$ m
At maximum depth of 8 ft (96 in.), $h > h_{int}$ At maximum depth of 2.5 m, $h > h_{int}$
then calculate L_2 (Eq 5-14).   then calculate L_2 (Eq 5-14).
$L_2 = [t_w + 72\ \text{in.} + LLDF\ (h)]/2$ $L_2 = [t_w + 1.83\ \text{m} + LLDF\ (h)]/2$
$L_2 = [20 + 72 + 1.15(96)]/2 = 101$ in. $L_2 = [.508 + 1.83 + 1.15(2.5)]/2 = 2.61$ m

At minimum depth of 4 ft,   At minimum depth of 1.2 m,

$W_L = \dfrac{1.2(20,000)(1.17)}{(65)(74)} = 5.84$ psi $W_L = \dfrac{1.2(90,000)(1.17)}{(1.63)(1.86)} = 41,768$ N/m^2

At maximum depth of 8 ft,   At maximum depth of 2.5 m,

$W_L = \dfrac{1.2(20,000)(1.0)}{(120)(101)} = 1.98$ psi $W_L = \dfrac{1.2(90,000)(1.0)}{(3.13)(2.61)} = 13,220$ N/m^2

Step 7. Calculate the composite constrained soil modulus (Eq 5-19).

Determine S_c from Table 5-5.

$M_{sn} = 10,000$ psi $M_{sn} = 69$ MPa
$M_{sb} = 1,125$ psi at 4 ft (saturated) $M_{sb} = 7.8$ MPa at 1.2 m (saturated)

BURIED PIPE DESIGN 71

M_{sb} = 1,219 psi at 8 ft (saturated) M_{sb} = 8.4 MPa at 2.5 m (saturated)
B_d = 54 in. at springline = 4.5 ft B_d = 1.35 m at springline
D = 36.7 in. = 3.06 ft D = 924 mm = 0.924 m
B_d/D = 4.5/3.06 = 1.47 B_d/D = 1.35/0.924 = 1.46

At 4 ft, M_{sn}/M_{sb} = 10,000/1,125 = 8.89   At 1.2 m, M_{sn}/M_{sb} = 69/7.8 = 8.88
S_c = 2.3 S_c = 2.3

At 8 ft, M_{sn}/M_{sb} = 10,000/1,219 = 8.2   At 2.5 m, M_{sn}/M_{sb} = 69/8.4 = 8.2
S_c = 2.3 S_c = 2.3
M_s = 1,125 (2.3) = 2,588 psi at 4 ft M_s = 7.8 (2.3) = 17.9 MPa at 1.2 m
M_s = 1,219 (2.3) = 2,804 psi at 8 ft M_s = 8.4 (2.3) = 19.3 MPa at 2.5 m

Step 8. Calculate the predicted deflection (Eq 5-8).

$$\frac{\Delta_y}{D} = \frac{(D_L W_c + W_L)K_x}{0.149 \text{PS} + 0.061 M_s} \qquad \frac{\Delta_y}{D} = \frac{(D_L W_c + W_L)K_x}{149 \text{PS} + 61000 M_s}$$

At 4 ft depth,   At 1.2 m depth,

$$= \frac{[1.05(3.3) + 5.84]0.1}{0.149(36) + 0.061(2,923)} \qquad = \frac{[1.05(22,560) + 41,768]0.1}{149(250) + 61,000(20.14)}$$

$= 0.0051$ $= 0.0052$
$= 0.51\%$ $= 0.52\%$

At 8 ft depth,   At 2.5 m depth,

$$= \frac{[1.05(6.66) + 1.98]0.1 \times 100}{0.149(36) + 0.061(2,962)} \qquad = \frac{[1.05(47,000) + 13,220]0.1}{149(250) + 61,000(20.5)}$$

$= 0.0048$ $= 0.0049$
$= 0.48\%$ $= 0.49\%$

Predicted deflection is less than the maximum permitted deflection of 5% and the maximum allowable deflection of 8.1%   Predicted deflection is less than the maximum permitted deflection of 5% and the maximum allowable deflection of 8.0%

∴ OK   ∴ OK

Step 9. Check combined loading (Eq 5-22 and 5-23).

$$\varepsilon_{pr} = \frac{P_w D}{2t(E_{HT})} \qquad \varepsilon_{pr} = \frac{P_w D}{2 \times 10^6 (t)(E_{HT})}$$

$$= \frac{115(36.69)}{2(.61)(1.8 \times 10^6)} \qquad = \frac{800(908.5)}{2 \times 10^6 (15.5)(12.5)}$$

$= 0.0019$ in./in. $= 0.0019$ mm/mm

72 FIBERGLASS PIPE DESIGN

Calculate ε_b (Eq 5-6).

$\varepsilon_b = (D_f)(\Delta y_a/D)(t_t/D)$
$= 5.5(0.05)(.65/36.69)$
$= 0.00487$ in./in.
$r_c = 1 - P_w/435 = 1 - 115/435 = 0.73$

$\varepsilon_b = (D_f)(\Delta y_a/D)(t_t/D)$
$= 5.5(0.05)(.65/36.69)$
$= 0.00487$ mm/mm
$r_c = 1 - P_w/3{,}000 = 1 - 800/3{,}000 = 0.73$

$$\frac{\varepsilon_{pr}}{\text{HDB}} \leq \frac{1-\left(\frac{\varepsilon_b r_c}{S_b}\right)}{FS_{pr}}$$

$$\frac{\varepsilon_{pr}}{\text{HDB}} \leq \frac{1-\left(\frac{\varepsilon_b r_c}{S_b}\right)}{FS_{pr}}$$

$$\frac{0.0019}{0.0065} \leq \frac{1-\left[\frac{(0.00487)(0.73)}{0.012}\right]}{1.8}$$

$$\frac{0.0019}{0.0065} \leq \frac{1-\left[\frac{(0.00487)(0.73)}{0.012}\right]}{1.8}$$

$0.292 \leq 0.39$ ∴ OK

$0.292 \leq 0.39$ ∴ OK

$$\frac{\varepsilon_b r_c}{S_b} \leq \frac{1-\left(\frac{\varepsilon_{pr}}{\text{HDB}}\right)}{FS_b}$$

$$\frac{\varepsilon_b r_c}{S_b} \leq \frac{1-\left(\frac{\varepsilon_{pr}}{\text{HDB}}\right)}{FS_b}$$

$$\frac{0.00487(0.73)}{0.012} \leq \frac{1-\left(\frac{0.0019}{0.0065}\right)}{1.5}$$

$$\frac{0.00487(0.73)}{0.012} \leq \frac{1-\left(\frac{0.0019}{0.0065}\right)}{1.5}$$

$0.296 \leq 0.47$ ∴ OK

$0.296 \leq 0.47$ ∴ OK

Step 10. Check buckling (Eq 5-24a).

$$q_a = \frac{(1.2 C_n)(EI)^{0.33}(\varphi_s M_s k_\upsilon)^{0.67} R_h}{(FS)r}$$

$$q_a = \frac{(1.2 C_n)(EI)^{0.33}(\varphi_s 10^6 M_s k_\upsilon)^{0.67} R_h}{(FS)r}$$

R_h = correction factor for depth of fill
$= 11.4/(11 + D/h)$

R_h = correction factor for depth of fill
$= 11.4/(11 + D/1{,}000h)$

At 4 ft (48 in.) depth,
$= 11.4/(11 + 36.69/48)$
$= 0.97$

At 1.2 m depth,
$= 11.4/(11 + 908.5/1{,}200)$
$= 0.97$

At 8 ft (96 in.) depth,
$= 11.4/(11 + 36.69/96)$
$= 1.00$

At 2.5 m depth,
$= 11.4/(11 + 908.5/2{,}500)$
$= 1.00$

$EI = PS(.149)[r + \Delta y/2]^3$
$\quad = 36(0.149)[36.69/2 + (.05)(36.69)/2]^3$

$\quad = 38{,}336$

At 4 ft depth,

$q_a = 1.2(0.55)(38{,}336)^{0.33}$
$\quad [0.9(2{,}588)(0.74)]^{0.67}(0.97)$
$\quad /2.5(36.69/2)$

$\quad = 69.4 \text{ psi}$

At 8 ft depth,

$q_a = 1.2(0.55)(38{,}336)^{0.33}$
$\quad [0.9(2{,}804)(0.74)]^{0.67}(0.97)$
$\quad /2.5(36.69/2)$

$\quad = 75.4 \text{ psi}$

$EI = PS \times 10^{-6}(.149)[r + \Delta y/2]^3$
$\quad = (250)(10^{-6})(0.149)[908.5/2$
$\quad\quad + .05(908.5)/2]^3$
$\quad = 4{,}043$

At 1.2 m depth,

$q_a = 1.2(0.55)(4{,}043)^{0.33}$
$\quad [0.9(17.9 \times 10^6)(0.74)]^{0.67}(0.97)$
$\quad /2.5(908.5)/2$

$\quad = 495 \text{ kPa}$

At 2.5 m depth,

$q_a = 1.2(0.55)(4{,}043)^{0.33}$
$\quad [0.9(19.3 \times 10^6)(0.74)]^{0.67}(0.97)$
$\quad /2.5(908.5)/2$

$\quad = 536 \text{ kPa}$

Check against requirement considering vacuum (Eq 5-25).

$\gamma_w h_w + R_w(W_c) + P_v \leq q_a$

R_w = water buoyancy factor
$\quad = 1 - 0.33(h_w/h) \ [0 \leq h_w \leq h]$

At 4 ft (48 in.) depth,
$\quad = 1 - 0.33 \ (12/48)$
$\quad = 0.918$

At 8 ft (96 in.) depth,
$\quad = 1 - 0.33 \ (60/96)$
$\quad = 0.794$

Buckling check at 4 ft depth:
$0.0361(12) + 0.918(3.33) + 14.7 \leq 69.4$
$18.19 \text{ psi} \leq 69.4 \text{ psi} \therefore \text{OK}$

Buckling check at 8 ft depth:
$0.0361(60) + 0.794(6.66) + 14.7 \leq 75.4$
$22.15 \text{ psi} \leq 75.4 \text{psi} \therefore \text{OK}$

$[\gamma_w h_w + R_w(W_c)] \times 10^{-3} + P_v \leq q_a$

R_w = water buoyancy factor
$\quad = 1 - 0.33(h_w/h) \ [0 \leq h_w \leq h]$

At 1.2 m depth,
$\quad = 1 - 0.33 \ (0.2/1.2)$
$\quad = 0.945$

At 2.5 m depth,
$\quad = 1 - 0.33 \ (1.5/2.5)$
$\quad = 0.802$

Buckling check at 1.2 m depth:
$[9{,}800(0.2) + 0.945(22{,}560)] \times 10^{-3}$
$+ 100 \leq 495$
$123 \text{ kPa} \leq 495 \text{ kPa} \therefore \text{OK}$

Buckling check at 2.5 m depth:
$[9{,}800(1.5) + 0.802(47{,}000)] \times 10^{-3}$
$+ 100 \leq 536$
$152 \text{ kPa} \leq 536 \text{ kPa} \therefore \text{OK}$

Check against requirement considering live load (Eq 5-26).

$$\gamma_w h_w + R_w(W_c) + W_L \leq q_a$$

$$[\gamma_w h_w + R_w(W_c) + W_L] \times 10^{-3} \leq q_a$$

Buckling check at 4 ft depth:
0.0361(12) + 0.918(3.33) + 5.84 ≤ 69.4
9.3 ≤ 69.4 ∴ OK

Buckling check at 1.2 m depth:
[9,800(0.2) + 0.945(22,560) + 41,768]
× 10⁻³ ≤ 495
65.0 kPa ≤ 495 kPa ∴ OK

Buckling check at 8 ft depth:
0.0361(60) + 0.794(6.66) + 1.98 ≤ 75.4
9.4 ≤ 75.4 ∴ OK

Buckling check at 2.5 m depth:
[9,800(1.5) + 0.802(47,000) + 13,220]
× 10⁻³ ≤ 536
65.6 kPa ≤ 536 kPa ∴ OK

REFERENCES

American Association of State Highway and Transportation Officials. 1999. *AASHTO LRFD Bridge Design Specifications*, 2nd ed. Washington, D.C.: American Association of State Highway and Transportation Officials.

American Society for Testing and Materials. ASTM D698, *Standard Test Methods for Laboratory Compaction Characteristics of Soil Using Standard Effort*. West Conshohocken, Pa.: American Society for Testing and Materials.

———. ASTM D1586, *Standard Test Method for Penetration Test and Split-Barrel Sampling of Soils*. West Conshohocken, Pa.: American Society for Testing and Materials.

———. ASTM D2412, *Standard Test Method for Determination of External Loading Characteristics of Plastic Pipe by Parallel-Plate Loading*. West Conshohocken, Pa.: American Society for Testing and Materials.

———. ASTM D2487, *Standard Classification of Soils for Engineering Purposes (Unified Soil Classification System)*. West Conshohocken, Pa.: American Society for Testing and Materials.

———. ASTM D3567, *Standard Practice for Determining Dimensions of "Fiberglass" (Glass-Fiber-Reinforced Thermosetting Resin) Pipe and Fittings*. West Conshohocken, Pa.: American Society for Testing and Materials.

———. ASTM D4253, *Standard Test Methods for Maximum Index Density and Unit Weight of Soils Using a Vibratory Table*. West Conshohocken, Pa.: American Society for Testing and Materials.

———. ASTM D4254, *Standard Test Methods for Minimum Index Density and Unit Weight of Soils and Calculation of Relative Density*. West Conshohocken, Pa.: American Society for Testing and Materials.

American Water Works Association. ANSI/AWWA C950, *Standard for Fiberglass Pressure Pipe*. Denver, Colo.: American Water Works Association.

Cagle, L., and B.C. Glascock. 1982. Recommended Design Requirements for Elastic Buckling of Buried Flexible Pipe (Report of ANSI/AWWA Standard C950 Ad-Hoc Task Group on Buckling). In *Proc. of AWWA Annual Conference* and SPI 39th Annual Conference (January 1984). Denver, Colo.: American Water Works Association.

Howard, A.K. 1977. Modulus of Soil Reaction Values for Buried Flexible Pipe. *Journal of Geotechnical Engineering*, 103:GTL.

———. 1996. *Pipeline Installation*. Lakewood, Colo.: Relativity Publishing.

Luscher, U. 1966. Buckling of Soil Surrounded Tubes. *Jour. Soil Mech. & Found.*, 92(6):213.

McGrath, T.J. 1998. Replacing E′ With the Constrained Modulus in Buried Pipe Design. In *Pipelines in the Constructed Environment*. Edited by J.P. Castronovo and J.A. Clark. Reston, Va.: American Society of Civil Engineers.

Molin, J. 1971. Principles of Calculation for Underground Plastic Pipes—Calculations of Loads, Deflection, Strain. *ISO Bull.*, 2(10):21.

Spangler, M.G., and R.L. Handy. 1982. *Soil Engineering*, 4th ed. New York: Harper & Row.

AWWA MANUAL M45

Chapter 6

Guidelines for Underground Installation of Fiberglass Pipe

6.1 INTRODUCTION

The structural and installation designs of fiberglass pipe, or almost any buried pipe, are closely related. The structural design process, discussed in chapter 5, assumes that a pipe will receive support from the surrounding soil, and the installation process must ensure that the support is provided. The guidelines in this chapter suggest procedures for burial of fiberglass pipe in typically encountered soil conditions. Recommendations for trenching, placing, and joining pipe; placing and compacting backfill; and monitoring deflection levels are included.

ANSI/AWWA Standard C950 specifies pipe that encompass a wide range of product variables. Diameters range from 1 in. to 12 ft (25 mm to 3,700 mm), pipe stiffnesses range from 9 psi to 72 psi (62 kPa to 496 kPa), and internal pressure ratings range up to 250 psi (1,700 kPa). Engineers and installers should recognize that all possible combinations of pipe, soil types, and natural ground conditions that may occur are not considered in this chapter. The recommendations provided may need to be modified or expanded to meet the needs of some installation conditions. Section 6.9 lists areas that may be influenced by project, local, or regional conditions and should be given consideration when preparing specifications. Guidance for installation of fiberglass pipe in subaqueous conditions is not included.

These guidelines are for use by engineers and specifiers, manufacturers, installation contractors, regulatory agencies, owners, and inspection organizations that are involved in the construction of buried fiberglass pipelines.

6.2 RELATED DOCUMENTS

The following ASTM standards provide engineers with additional information related to installing buried pipe.

D8	Standard Terminology Relating to Materials for Roads and Pavements
D420	Standard Guide to Site Characterization for Engineering, Design, and Construction Purposes
D653	Standard Terminology Relating to Soil, Rock, and Contained Fluids
D698	Standard Test Methods for Laboratory Compaction Characteristics of Soil Using Standard Effort (12,400 ft-lbf/ft^3 [600 kN-m/m^3])
D883	Standard Terminology Relating to Plastics
D1556	Standard Test Method for Density and Unit Weight of Soil in Place by the Sand-Cone Method
D1557	Standard Test Methods for Laboratory Compaction Characteristics of Soil Using Modified Effort (56,000 ft-lbf/ft^3 [2,700 kN-m/m^3])
D1586	Standard Test Method for Penetration Test and Split-Barrel Sampling of Soils
D2167	Standard Test Method for Density and Unit Weight of Soil in Place by the Rubber Balloon Method
D2216	Standard Test Method for Laboratory Determination of Water (Moisture) Content of Soil and Rock by Mass
D2321	Standard Practice for Underground Installation of Thermoplastic Pipe for Sewers and Other Gravity-Flow Applications
D2412	Standard Test Method for Determination of External Loading Characteristics of Plastic Pipe by Parallel-Plate Loading
D2487	Standard Classification of Soils for Engineering Purposes (Unified Soil Classification System)
D2488	Standard Practice for Description and Identification of Soils (Visual–Manual Procedure)
D2922	Standard Test Methods for Density of Soil and Soil-Aggregate in Place by Nuclear Methods (Shallow Depth)
D3017	Standard Test Method for Water Content of Soil and Rock in Place by Nuclear Methods (Shallow Depth)
D3441	Standard Test Method for Mechanical Cone Penetration Tests of Soil
D3839	Standard Guide for Underground Installation of "Fiberglass" (Glass-Fiber-Reinforced Thermosetting-Resin) Pipe
D4253	Standard Test Methods for Maximum Index Density and Unit Weight of Soils Using a Vibratory Table
D4254	Standard Test Methods for Minimum Index Density and Unit Weight of Soils and Calculation of Relative Density
D4318	Standard Test Methods for Liquid Limit, Plastic Limit, and Plasticity Index of Soils

D4564 Standard Test Method for Density of Soil in Place by the Sleeve Method

D4643 Standard Test Method for Determination of Water (Moisture) Content of Soil by the Microwave Oven Method

D4914 Standard Test Methods for Density of Soil and Rock in Place by the Sand Replacement Method in a Test Pit

D4944 Standard Test Method for Field Determination of Water (Moisture) Content of Soil by the Calcium Carbide Gas Pressure Tester

D4959 Standard Test Method for Determination of Water (Moisture) Content of Soil by Direct Heating

D5030 Standard Test Method for Density of Soil and Rock in Place by the Water Replacement Method in a Test Pit

D5080 Standard Test Method for Rapid Determination of Percent Compaction

F412 Standard Terminology Relating to Plastic Piping Systems

F1668 Standard Guide for Construction Procedures for Buried Plastic Pipe

6.3 TERMINOLOGY

Terminology used in this chapter is in accordance with ASTM Standards D8, D653, D883, and F412 unless otherwise indicated. The following terms are specific to this manual:

Bedding. Backfill material placed in the bottom of the trench or on the foundation to provide a uniform material on which to lay the pipe; the bedding may or may not include part of the haunch zone (see Figure 6-1).

Compactibility. A measure of the ease with which a soil may be compacted to a high density and high stiffness. Crushed rock has high compactibility because a dense and stiff state may be achieved with little compactive energy.

Deflection. Any change in the diameter of the pipe resulting from installation and imposed loads. Deflection may be measured and reported as change in either vertical or horizontal diameter and is usually expressed as a percentage of the undeflected pipe diameter.

Engineer. The engineer or the duly recognized or authorized representative in responsible charge of the work.

Final backfill. Backfill material placed from the top of the initial backfill to the ground surface (see Figure 6-1).

Fines. Soil particles that pass a No. 200 (0.076-mm) sieve.

Foundation. Material placed and compacted in the bottom of the trench to replace overexcavated material and/or to stabilize the trench bottom if unsuitable ground conditions are encountered (see Figure 6-1).

Geotextile. Any permeable textile material used with foundation, soil, earth, rock, or any other geotechnical engineering-related material as an integral part of a synthetic product, structure, or system.

Haunching. Backfill material placed on top of the bedding and under the springline of the pipe; the term only pertains to soil directly beneath the pipe (see Figure 6-1).

Initial backfill. Backfill material placed at the sides of the pipe and up to 6 in. to 12 in. (150 mm to 300 mm) over the top of the pipe, including the haunching (see Figure 6-1).

78   FIBERGLASS PIPE DESIGN

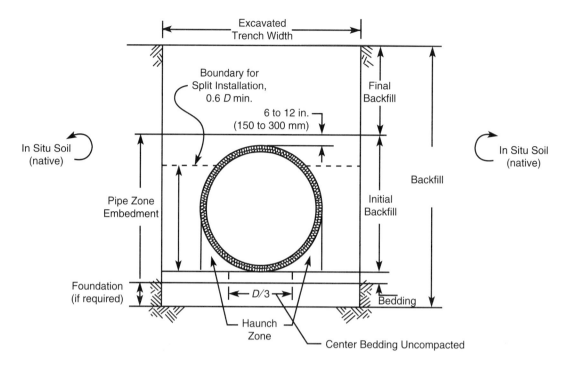

Figure 6-1   Trench cross-section terminology

Manufactured aggregates. Aggregates such as slag that are products or by-products of a manufacturing process, or natural aggregates that are reduced to their final form by a manufacturing process such as crushing.

Maximum standard Proctor density. The maximum dry density of soil compacted at optimum moisture content and with standard effort in accordance with ASTM D698.

Native (in situ) soil. Natural soil in which a trench is excavated for pipe installation or on which a pipe and embankment are placed.

Open-graded aggregate. An aggregate that has a particle size distribution such that when compacted, the resulting voids between the aggregate particles are relatively large.

Optimum moisture content. The moisture content of soil at which its maximum density is obtained (see ASTM D698 and ASTM D1557).

Pipe zone embedment. All backfill around the pipe, including the bedding, haunching, and initial backfill.

Processed aggregates. Aggregates that are screened, washed, mixed, or blended to produce a specific particle size distribution.

Relative density. A measure of the density of a granular soil "relative" to the soil in its loosest state and the soil in its densest state (see ASTM D653 for a precise definition), as obtained by laboratory testing in accordance with ASTM D4253 and ASTM D4254.

Soil stiffness. A property of soil, generally represented numerically by a modulus of deformation, that indicates the relative amount of deformation that will occur under a given load.

Split installation. An installation where the initial backfill is composed of two different materials or one material placed at two different densities. The lower material extends from the top of the bedding to a depth of at least 0.6 times the diameter.

6.4 IN SITU SOILS

It is important to understand in situ conditions prior to construction in order to prepare proper specifications and plan construction methods. Classification of soils according to ASTM D2487 and ASTM D2488 is useful in gaining an understanding of in situ conditions. Other tests, such as the standard penetration and cone penetrometer tests, are also useful in determining soil stiffness. Depending on actual installation conditions, such as trench geometry, the in situ soil conditions may also have a significant impact on pipe design. Refer to chapter 5 for further discussion.

Consideration should also be given to seasonal or long-term variations in groundwater level when evaluating groundwater conditions. For example, if the soil exploration program is conducted in August, the groundwater level may be quite low compared to levels in April or May.

6.5 EMBEDMENT MATERIALS

Soil types used or encountered when burying pipes include those classified in Table 5-2 as well as natural, manufactured, and processed aggregates. The soil classifications are grouped into soil "stiffness categories" (SC#) in Table 6-1, based on the typical soil stiffness when compacted. Soil SC1 indicates a soil with high compactibility, i.e., a soil that provides the highest stiffness at any given percentage of maximum Proctor density and a soil that provides a given stiffness with the least compactive energy. Each higher number soil stiffness category is successively less compactible, i.e., it provides less soil stiffness at a given percentage of maximum Proctor density and requires greater compactive energy to provide a given level of stiffness. See chapter 5 for a discussion of how soil stiffness affects buried pipe behavior.

Table 6-2 provides recommendations on installation and use of embedment materials based on stiffness category and location in the trench. In general, soil conforming to SC1 through SC4 should be used as recommended and SC5 materials should be excluded from the pipe zone embedment.

6.5.1 Soil Stiffness Classes

Soil stiffness category 1 (SC1). SC1 materials provide maximum pipe support for a given percent compaction due to low content of sand and fines. With minimum effort these materials can be installed at relatively high soil stiffnesses over a wide range of moisture contents. In addition, the high permeability of SC1 materials may aid in the control of water, making them desirable for embedment in rock cuts where water is frequently encountered. However, when groundwater flow is anticipated, consideration should be given to the potential for migration of fines from adjacent materials into the open-graded SC1 material (see Sec. 6.5.2).

Soil stiffness category 2 (SC2). When compacted, SC2 materials provide a relatively high level of pipe support. However, open-graded groups may allow migration and the sizes should be checked for compatibility with adjacent material (see Sec. 6.5.2).

Soil stiffness category 3 (SC3). SC3 materials provide less support for a given density than SC1 or SC2 materials. Higher levels of compactive effort are required and moisture content must be near optimum to minimize compactive effort and achieve the required density. These materials provide reasonable levels of pipe support once proper density is achieved.

Soil stiffness category 4 (SC4). SC4 materials require a geotechnical evaluation prior to use. The moisture content must be near optimum to minimize compactive

Table 6-1 Soil stiffness categories

Soil Stiffness Category	Unified Soil Classification System Soil Groups (note 1)	American Association of State Highway and Transportation Officials (AASHTO) Soil Groups (note 2)
SC1	Crushed rock: ≤15% sand, maximum 25% passing the 3/8-in. sieve and maximum 5% passing No. 200 sieve (note 3)	
SC2	Clean, coarse-grained soils: SW, SP, GW, GP, or any soil beginning with one of these symbols with 12% or less passing No. 200 sieve (note 4)	A1, A3
SC3	Coarse-grained soils with fines: GM, GC, SM, SC, or any soil beginning with one of these symbols with more than 12% fines	A-2-4, A-2-5, A-2-6, or A-4 or A-6 soils with more than 30% retained on a No. 200 sieve
	Sandy or gravelly fine-grained soils: CL, ML (or CL-ML, CL/ML, ML/CL) with more than 30% retained on a No. 200 sieve	
SC4	Fine-grained soils: CL, ML (or CL-ML, CL/ML, ML/CL) with 30% or less retained on a No. 200 sieve	A-2-7, or A-4 or A-6 soils with 30% or less retained on a No. 200 sieve
SC5	Highly plastic and organic soils: MH, CH, OL, OH, PT	A5, A7

NOTES:

1. ASTM D2487, *Standard Classification of Soils for Engineering Purposes* (Unified Soil Classification System).
2. AASHTO M145, *Classification of Soils and Soil Aggregate Mixtures.*
3. SC1 soils have higher stiffness than SC2 soils, but data on specific soil stiffness values is not available at the current time. Until such data is available, the soil stiffness of dumped SC1 soils can be taken to be equivalent to SC2 soils compacted to 90% of maximum standard Proctor density (SC2-90), and the stiffness of compacted SC1 soils can be taken to be equivalent to SC2 soils compacted to 100% of maximum standard Proctor density (SC2-100). Even if dumped, SC1 materials should always be worked into the haunch zone (see Sec. 6.7.3).
4. Uniform fine sands (SP) with more than 50% passing a No. 100 sieve (0.006 in., 0.15 mm) are very sensitive to moisture and should not be used as backfill for fiberglass pipe unless specifically allowed in the contract documents. If use of these materials is allowed, compaction and handling procedures should follow the guidelines for SC3 materials.

effort and achieve the required density. When properly placed and compacted, SC4 materials can provide reasonable levels of pipe support; however, these materials may not be suitable under high fills, surface-applied wheel loads, or high–energy-level vibratory compactors and tampers. Do not use where water conditions in the trench prevent proper placement and compaction.

Soil stiffness category 5 (SC5). SC5 materials are not suitable for use as backfill for flexible pipe and must be excluded from the pipe zone embedment.

6.5.2 Considerations for Use of Soil in Backfill

Moisture content of embedment materials. The moisture content of embedment materials with substantial fines must be controlled to permit placement and compaction to required levels. For soils with low permeability (i.e., SC3, SC4, and some dual-symbol SC2 soils), moisture content is typically controlled to ±3% of optimum (ASTM D698). Obtaining and maintaining the required limits on moisture content are important criteria for selecting materials, because failure to achieve required density, especially in the pipe zone embedment, may result in excessive deflection.

Table 6-2 Recommendations for installation and use of soils and aggregates for foundation and pipe zone embedment

Soil Stiffness Category*	SC1	SC2	SC3	SC4
General recommendations and restrictions	Acceptable and common where no migration is probable or when combined with a geotextile filter media. Suitable for use as a drainage blanket and underdrain where adjacent material is suitably graded or when used with a geotextile filter fabric (see Sec. 6.5.2).	Where hydraulic gradient exists, check gradation to minimize migration. Clean groups are suitable for use as a drainage blanket and underdrain (see Table 5-3). Uniform fine sands (SP) with more than 50% passing a No. 100 sieve (0.006 in., 0.15 mm) behave like silts and should be treated as SC3 soils.	Do not use where water conditions in trench prevent proper placement and compaction. Not recommended for use with pipes with stiffness of 9 psi (62 kPa) or less.	Difficult to achieve high soil stiffness (see Sec. 6.5.1). Do not use where water conditions in trench prevent proper placement and compaction. Not recommended for use with pipes with stiffness of 9 psi (62 kPa) or less.
Foundation	Suitable as foundation and for replacing overexcavated and unstable trench bottom as restricted above.	Suitable as foundation and for replacing overexcavated and unstable trench bottom as restricted above. Install and compact in 12-in. (300-mm) maximum layers.	Suitable for replacing overexcavated trench bottom as restricted above. Install and compact in 6-in. (150-mm) maximum layers.	Not suitable.
Pipe zone embedment	Suitable as restricted above. Work material under pipe to provide uniform haunch support.	Suitable as restricted above. Work material under pipe to provide uniform haunch support.	Suitable as restricted above. Difficult to place and compact in the haunch zone.	Suitable as restricted above. Difficult to place and compact in the haunch zone.
Embedment compaction				
Minimum recommended density, SPD†	Minimum density typically achieved by dumped placement.	85%	90%	95%
Relative compactive effort required to achieve minimum density	Low	Moderate	High	Very high
Compaction methods	Vibration or impact	Vibration or impact	Impact	Impact
Required moisture control	None	None	Maintain near optimum to minimize compactive effort.	Maintain near optimum to minimize compactive effort.

* SC5 materials are unsuitable as embedment. They may be used as final backfill as permitted by the engineer.
† SPD is standard Proctor density as determined by ASTM Test Method D698 (AASHTO T-99).

Table 6-3 Maximum particle size for pipe embedment

Nominal Diameter (D_i) Range		Maximum Particle Size	
in.	mm	in.	mm
$D_i \leq 18$	$D_i \leq 450$	0.5	13
$18 < D_i \leq 24$	$450 < D_i \leq 600$	0.75	19
$24 < D_i \leq 36$	$600 < D_i \leq 900$	1.00	25
$36 < D_i \leq 48$	$900 < D_i \leq 1,200$	1.25	32
$48 < D_i$	$1,200 < D_i$	1.50	38

SC1 and most SC2 materials are free draining and require little or no control of moisture for compaction.

Compatibility of pipe and backfill. Experience has shown that pipe deflections and strain levels increase when low-stiffness pipe is embedded in backfill materials that require large compactive efforts. This occurs because of the local distortions of the pipe shape that result as compactive energy is applied to the backfill. Because of this, it is recommended that pipe with stiffness of 9 psi (62 kPa) or less be embedded only in soil types SC1 or SC2.

Maximum particle size. Maximum particle size for pipe zone embedment is limited based on pipe diameter, as listed in Table 6-3. For final backfill, the maximum particle size allowed should not exceed 75% of the lift thickness. When final backfill contains cobbles, boulders, etc., the initial bedding should be extended above the top of the pipe at least 12 in. (300 mm). Backfill containing boulders larger than 8 in. (200 mm) shall not be dropped or rolled down sloping trench walls onto the backfill from a height greater than 6 ft (1.8 m) until the depth of fill over the top of the pipe is greater than 24 in. (600 mm).

Migration. When open-graded material is placed adjacent to finer material, fines may migrate into the coarser material under the action of hydraulic gradient from groundwater flow. Significant hydraulic gradients may arise in the pipeline trench during construction, when water levels are controlled by various pumping or well-pointing methods, or after construction, when permeable underdrain or embedment materials act as a "french" drain under high groundwater levels. Field experience shows that migration can result in significant loss of pipe support and increasing deflections that may eventually exceed design limits. The gradation and relative size of the embedment and adjacent materials must be compatible in order to minimize migration. In general, where significant groundwater flow is anticipated, avoid placing coarse, open-graded materials, such as SC1, above, below, or adjacent to finer materials, unless methods are employed to impede migration. For example, consider the use of an appropriate soil filter or a geotextile filter fabric along the boundary of the incompatible materials.

The following filter gradation criteria may be used to restrict migration of fines into the voids of coarser material under a hydraulic gradient:

- $D_{15}/d_{85} < 5$ where D_{15} is the sieve opening size passing 15% by weight of the coarser material and d_{85} is the sieve opening size passing 85% by weight of the finer material.

- $D_{50}/d_{50} < 25$ where D_{50} is the sieve opening size passing 50% by weight of the coarser material and d_{50} is the sieve opening size passing 50% by weight of the finer material. This criterion need not apply if the coarser material is well graded (see ASTM D2487).

If the finer material is a medium to highly plastic clay (CL or CH), the following criterion may be used in lieu of the D_{15}/d_{85} criteria: $D_{15} < 0.02$ in. (0.5 mm) where D_{15} is the sieve opening size passing 15% by weight of the coarser material.

The aforementioned criteria may need to be modified if one of the materials is gap graded. Materials selected for use based on filter gradation criteria should be handled and placed in a manner that will minimize segregation.

Cementitious backfill materials. Backfill materials supplemented with cement to improve long-term strength and/or stiffness (soil cement, cement-stabilized backfill) or to improve flowability (flowable fill, controlled low-strength material) have been shown to be effective backfill materials in terms of ease of placement and quality of support to pipe. Although not specifically addressed by this manual, use of these materials is beneficial under many circumstances.

6.6 TRENCH EXCAVATION

6.6.1 Excavation

Excavate trenches to ensure that sides will be stable under all working conditions. Slope trench walls or provide supports in conformance with safety standards. Open only enough trench that can be safely maintained by available equipment. Place and compact backfill in trenches as soon as practicable, preferably no later than the end of each working day. Place excavated material away from the edge of the trench to minimize the risk of trench wall collapse.

Water control. It is always good practice to remove water from a trench before laying and backfilling pipe. Although circumstances occasionally require pipe installation in conditions of standing or running water, such practice is outside the scope of this chapter. Prevent runoff and surface water from entering the trench at all times.

Groundwater. When groundwater is present in the work area, dewater to maintain stability of in situ and imported materials. Maintain water level below pipe bedding. Use sump pumps, well points, deep wells, geotextiles, perforated underdrains, or stone blankets of sufficient thickness to remove and control water in the trench. When excavating, ensure the groundwater is below the bottom of the cut at all times to prevent washout from behind sheeting or sloughing of exposed trench walls. Maintain control of water in the trench before, during, and after pipe installation and until embedment is installed and sufficient backfill has been placed to prevent flotation of the pipe (see Sec. 6.7.3). To preclude loss of soil support, employ dewatering methods that minimize removal of fines and the creation of voids within in situ materials.

Running water. Control running water that emanates from surface drainage or groundwater to preclude undermining of the trench bottom or walls, the foundation, or other zones of embedment. Provide dams, cutoffs, or other barriers at regular intervals along the installation to preclude transport of water along the trench bottom. Backfill all trenches as soon as practical after the pipe is installed to prevent disturbance of pipe and embedment.

Materials for water control. Use suitably graded materials for foundation layers to transport running water to sump pits or other drains. Use properly graded materials and/or perforated underdrains to enhance transport of running water, as required. Select the gradation of the drainage materials to minimize migration of fines from surrounding materials (see Sec. 6.5.2).

Minimum trench width. Where trench walls are stable or supported, provide a width sufficient, but no greater than necessary, to ensure working room to properly and safely place and compact haunching and other embedment materials. The space between the pipe and trench wall must be 6 in. (150 mm) wider than the compaction equipment used in this region. For a single pipe in a trench, minimum width at the bottom of the trench should be 1.25 times the outside diameter of the pipe plus 12 in. (300 mm). For multiple pipes in the same trench, clear space between pipes must be at least the average of the radii of the two adjacent pipes for depths greater than 12 ft (3.5 m) and two thirds of the average of the radii of the two adjacent pipes for depths less than 12 ft (3.5 m). The distance from the outside pipe to the trench wall must not be less than if that pipe were installed as a single pipe in a trench. If mechanical compaction equipment is used, the minimum space between pipe and trench wall or between adjacent pipe shall not be less than the width of the widest piece of equipment plus 6 in. (150 mm).

In addition to safety considerations, the trench width in unsupported, unstable soils will depend on the size and stiffness of the pipe, stiffness of the embedment and in situ soil, and depth of cover. Specially designed equipment or the use of free-flowing backfill, such as uniform rounded pea gravel or flowable fill, may enable the satisfactory installation and embedment of pipe in trenches narrower than specified earlier. If the use of such equipment or backfill material provides an installation consistent with the requirements of this manual, minimum trench widths may be reduced if approved by the engineer.

Support of trench walls. When supports such as trench sheeting, trench jacks, or trench shields or boxes are used, ensure that support of the pipe embedment is maintained throughout the installation process. Ensure that sheeting is sufficiently tight to prevent washing out of the trench wall from behind the sheeting. Provide tight support of trench walls below viaducts, existing utilities, or other obstructions that restrict driving of sheeting.

Supports left in place. Sheeting driven into or below the top of the pipe zone should be left in place to preclude loss of support of foundation and embedment materials. When top of sheeting is to be cut off, make the cut 1.5 ft (0.5 m) or more above the crown of the pipe. Leave walers and braces in place as required to support cutoff sheeting and the trench wall in the vicinity of the pipe zone. Timber sheeting to be left in place is considered a permanent structural member and should be treated against biological degradation (e.g., attack by insects or other biological forms), as necessary, and against decay if above groundwater. Note that certain preservative and protective compounds may pose environmental hazards. Determination of acceptable compounds is outside the scope of this manual.

Movable trench wall supports. Do not disturb the installed pipe or the embedment when using movable trench boxes and shields. Movable supports should not be used below the top of the pipe embedment zone, unless approved methods are used for maintaining the integrity of embedment material. Before moving supports, place and compact embedment to sufficient depths to ensure protection of the pipe. As supports are moved, finish placing and compacting embedment.

Removal of trench wall support. If the removal of sheeting or other trench wall supports that extend below the top of the pipe is permitted, ensure that neither pipe, foundation, nor embedment materials are disturbed by support removal. Fill voids left after removal of supports and compact all material to required densities. Pulling the trench wall support in stages as backfilling progresses is advised.

6.6.2 Trench Bottom

Excavate trenches to specified grades. See Sec. 6.7.1 for guidance on installing foundation and bedding.

Excavate trench a minimum of 4 in. (100 mm) below the bottom of the pipe. When ledge, rock, hardpan, or other unyielding material or cobbles, rubble, debris, boulders, or stones larger than 1.5 in. (40 mm) are encountered in the trench bottom, excavate a minimum depth of 6 in. (150 mm) below the pipe bottom.

If the trench bottom is unstable or shows a "quick" tendency, overexcavate as required to provide the proper foundation.

The native material may be used for bedding and initial backfill if it meets all of the criteria of the specified pipe zone embedment materials. Trench preparation is discussed in Sec. 6.7.1.

6.6.3 Trenching on Slopes

The angle at which slopes can become unstable depends on the quality of the soil. The risk of unstable conditions increases dramatically with slope angle. In general, pipes should not be installed on slopes greater than 15° (a slope of 1:4) or in areas where slope instability is suspected, unless supporting conditions have been verified by a proper geotechnical investigation. Installing pipes aboveground may be a preferred method for steep slopes, because aboveground structures such as pipe supports are more easily defined and, therefore, the quality of installation is easier to monitor and settlement easier to detect. Pipes may be installed on slopes greater than 15° (a slope of 1:4) provided

- Long-term stability of the installation can be ensured with proper geotechnical design.

- Pipes are backfilled with coarse-grained material (SC1) with high shear strength or the shear strength of the backfill is assured by other means. The backfill should be compacted to at least 90% of maximum standard Proctor density (ASTM D698).

- Pipes are installed in straight alignment (±0.2°) with minimum gap between pipe ends.

- Absolute long-term movement of the backfill in the axial direction of the pipe is less than 0.75 in. (20 mm) to avoid joint separation.

- The installation is properly drained to avoid washout of materials and ensure adequate soil shear strength. This may include treatment in the backfill or on the ground surface.

- Stability of individual pipes is monitored throughout the construction phase and the first stages of operation.

- The manufacturer is consulted to determine if a special pipe design is required.

6.7 PIPE INSTALLATION

Recommendations for use of the various types of materials classified in Sec. 6.5.1 and Table 5-2 for foundation, bedding, haunching, and backfill are provided in Table 6-2. Installation of pipe in areas where significant settlement may be anticipated, such as

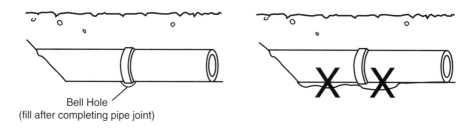

a. Proper Bedding Support **b. Improper Bedding Support**

Source: Flowtite Technology, Sandefjord, Norway.

Figure 6-2 Examples of bedding support

in backfill adjacent to building foundations, sanitary landfills, or in other highly unstable soils, requires special engineering and is outside the scope of this manual.

6.7.1 Preparation of Trench

Foundation and bedding. Provide a firm, stable, and uniform support for the pipe barrel and any protruding features of its joint (see Figure 6-2). Provide a minimum of 4 in. (100 mm) of bedding below the barrel and 3 in. (75 mm) below any part of the pipe, such as expanded bells, unless otherwise specified.

Bedding material. In general, the bedding material will need to be an imported material to provide the proper gradation and pipe support. It is preferable that the same material be used for the initial backfill. To determine if the native material is acceptable as a bedding material, it should meet all of the requirements of the initial backfill. This determination must be made constantly during the pipe installation process because native soil conditions vary widely and change suddenly along the length of a pipeline. It is becoming common practice to leave the bedding uncompacted for a width of one third of the pipe diameter centered directly under the pipe. This reduces concentrated loads on the invert (see Figure 6-1).

Rock and unyielding materials. When rock or unyielding material is present in the trench bottom, install a cushion of bedding, 6 in. (150 mm) minimum thickness, below the bottom of the pipe. If there is a sudden transition from rock to a softer material under the pipe, steps must be taken to accommodate possible differential settlement. Figure 6-3(b) illustrates one method; however, other methods are also possible.

Unstable trench bottom. Where the trench bottom is overexcavated because of unstable or "quick" conditions, install a foundation of SC1 or SC2 material. Use a suitably graded material where conditions may cause migration of fines and loss of pipe support. Place and compact foundation material in accordance with Table 6-2. For severe conditions, a special foundation, such as piles or sheeting capped with a concrete mat, may be required. The use of appropriate geotextiles can control quick and unstable trench bottom conditions.

Localized loadings. Minimize localized loadings and differential settlement wherever the pipe crosses other utilities or subsurface structures (see Figures 6-3 and 6-4) or whenever there are special foundations, such as concrete-capped piles or sheeting. Provide a 12-in. (300-mm) minimum cushion of bedding or compacted backfill between the pipe and any point of localized loading.

GUIDELINES FOR UNDERGROUND INSTALLATION OF FIBERGLASS PIPE 87

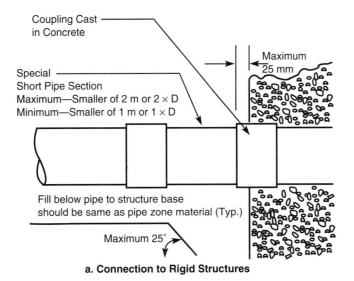

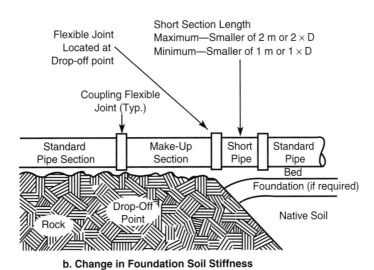

Source: Flowtite Technology, Sandefjord, Norway.

Figure 6-3 Accommodating differential settlement

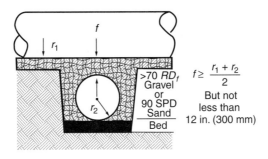

Source: Flowtite Technology, Sandefjord, Norway.

Figure 6-4 Cross-over of adjacent piping systems

Overexcavation. If the trench bottom is excavated below intended grade, fill the overexcavation with compatible foundation or bedding material and compact to a density not less than the minimum densities listed in Table 6-2.

Sloughing. If trench sidewalls slough off during any excavation or installation of pipe zone embedment, remove all sloughed and loose material from the trench.

6.7.2 Placing and Joining Pipe

Location and alignment. Place pipe and fittings in the trench with the invert conforming to the required elevations, slopes, and alignment. Provide bell holes in pipe bedding, no larger than necessary, in order to ensure uniform pipe support. Fill all voids under the bell by working in bedding material. In special cases where the pipe is to be installed to a curved alignment, maintain angular "joint deflection" (axial alignment) and pipe bending radius within acceptable design limits. Pipe should be laid on flat, uniform material that is at the appropriate grade. Do not bring pipe to grade by using mounds of soil or other material at discreet points along the length of the pipe. When pipe laying is interrupted, secure piping against movement and seal open ends to prevent the entrance of water, mud, or foreign material.

Jointing. Comply with manufacturer's recommendations for assembly of joint components, lubrication, and making of joints.

Elastomeric seal (gasketed) joints. Mark pipe ends, or verify that pipe ends are marked, to indicate insertion stop position and that pipe is inserted into pipe or fitting bells to this mark. Push spigot into bell using methods recommended by the manufacturer, keeping pipe true to line and grade. Protect the end of the pipe during homing and do not use excessive force that may result in overassembled joints or dislodged gaskets. If full entry is not achieved, disassemble and clean the joint and reassemble. Use only lubricant supplied or recommended for use by the pipe manufacturer. Do not exceed manufacturer's recommendations for angular "deflection" (axial alignment).

Adhesive bonded and wrapped joints. When making adhesive bonded and wrapped joints, follow recommendations of the pipe manufacturer. Allow freshly made joints to set for the recommended time before moving, burying, or otherwise disturbing the pipe.

Angularly deflected joints. Large radius bends in pipelines may be accomplished by rotating the alignment of adjacent lengths of pipe (i.e., "angularly deflecting" the joint). The amount of angular deflection should not exceed the manufacturer's recommendations.

6.7.3 Placing and Compacting Pipe Backfill Materials

Place embedment materials by methods that will not disturb or damage the pipe. Work in and compact the haunching material in the area between the bedding and the underside of the pipe before placing and compacting the remainder of the pipe zone embedment (see Figure 6-5). Do not permit compaction equipment to contact and damage the pipe. Use compaction equipment and techniques that are compatible with materials used and located in the trench.

Compaction of soils containing few fines (SC1 and SC2 with less than 5% fines). If compaction is required, use surface plate vibrators, vibratory rollers, or internal vibrators. The compacted lift thickness should not exceed 12 in. (300 mm) when compacted with surface plate vibrators or vibratory rollers; when compacted with internal vibrators, it should not exceed the length of the internal vibrators. Density determination should typically be in accordance with ASTM D4253 and ASTM

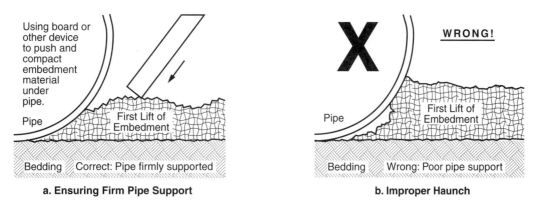

Source: Flowtite Technology, Sandefjord, Norway.

Figure 6-5 Proper compaction under haunches

D4254 (relative density). In some cases, the density of SW or SP soils may be determined by ASTM D698 (standard Proctor) if the test results in a clearly defined compaction curve.

Compaction of soils containing some fines (SC2 with 5 to 12% fines). These soils may behave as a soil containing few fines or as a soil containing a significant amount of fines. The methods of compaction and density determination should be based on the method that results in the higher in-place density.

Compaction of soils containing a significant amount of fines (SC3, SC4, and SC5). These soils should be compacted with impact tampers or sheepsfoot rollers. Density determination should be in accordance with ASTM D698 (standard Proctor). The maximum density occurs at the optimum moisture content. Less effort is required to reach a given density when the moisture content is within 2 percentage points of the optimum moisture. A rapid method of determining the percent compaction and moisture variation is described in ASTM D5080. For compaction levels of 90% standard Proctor or higher, the compacted lift thickness should not exceed 6 in. (150 mm).

Determination of the in-place density of soils. The in-place density of any in situ or fill soil may be determined in accordance with ASTM D1556, ASTM D2167, ASTM D2922, ASTM D4564, ASTM D4914, or ASTM D5030. The applicable test method will depend on the type of soil, moisture content of the soil, and the maximum particle size present in the soil. The moisture content of the soil may be determined in accordance with ASTM D2216, ASTM D3017, ASTM D4643, ASTM D4944, or ASTM D4959. When using nuclear density-moisture gauges (ASTM D2922 and ASTM D3017), the gauge should be site-calibrated in the proximity of the pipe and in the excavation as recommended by the gauge manufacturer.

Minimum density. The minimum embedment density should be established based on an evaluation of specific project conditions. Densities higher than those recommended in Table 6-2 may be appropriate. Minimum densities given in Table 6-2 are intended to provide satisfactory embedment stiffness in most installation conditions.

Densification using water. Densification of pipe zone embedment using water (jetting or saturation with vibration) should be done only under controlled conditions.

Backfill around angularly deflected pipe joints. When pipe joints are angularly rotated to accomplish large radii bends in pipelines that will operate at internal pressures of 15 psi (100 kPa) or greater, the backfill surrounding the joint should be compacted to at least 90% of maximum standard Proctor density for SC1 and SC2 materials and 95% of maximum standard Proctor density for SC3 and SC4 materials.

Consult the manufacturer for minimum depths of burial and additional restraint that may be required when the angular deflection is vertical.

Minimum cover. To preclude damage to the pipe and disturbance to pipe embedment, a minimum depth of backfill above the pipe should be maintained before allowing vehicles or heavy construction equipment to traverse the pipe trench. The minimum depth of cover should be established based on an evaluation of specific project conditions, such as pipe diameter and stiffness, soil type and stiffness, and live load type and magnitude. In the absence of an engineering evaluation, the following minimum cover requirements should be used.

For embedment materials installed to the minimum densities given in Table 6-2 and live loads similar to AASHTO H-20, provide cover (i.e., depth of backfill above top of pipe) of at least 24 in. (0.6 m) for SC1 embedment and cover of at least 36 in. (0.9 m) for SC2, SC3, or SC4 embedment before allowing vehicles or construction equipment to traverse the trench surface; provide at least 48 in. (1.2 m) of cover before using a hydrohammer for compaction. Where construction loads may be excessive (e.g., cranes, earth-moving equipment, or other vehicles with wheel loads that exceed the AASHTO H-20 loading), minimum cover should be increased or special structures, such as relief slabs at grade, may be installed to reduce the load transferred to the pipe.

If there is a risk of pipe flotation, the minimum cover should be 1 pipe diameter. If a specific analysis is made of the buoyant force of an empty pipe compared to the submerged weight of soil over the pipe, this minimum cover may be reduced.

6.7.4 Connections and Appurtenant Structures

Connections to manholes and rigid structures and changing foundation soils. When differential settlement can be expected, such as at the ends of casing pipe, when the pipe enters a manhole, at anchor blocks, or where foundation soils change stiffness, provide a flexible system capable of accommodating the anticipated settlement. This may be accomplished by placing a joint as close as practically possible to the face of the structure and a second joint within 1 to 2 pipe diameters of the face of the structure (see Figure 6-3). The short length of pipe, called a rocker pipe, shall be installed in straight alignment with the short pipe section coming out of the rigid structure. The rocker pipe should have a minimum pipe stiffness of 36 psi (248 kPa) to transition between lower stiffness pipe and the rigid structure. Multiple rocker pipes should not be used. Alternatively, attach the pipe to the rigid structure with a flexible boot capable of accommodating the anticipated differential movement. Extra care and caution must be taken to replace and *properly compact* backfill adjacent to any rigid structure. Construction of concrete structures will frequently require overexcavation for formwork, etc. This extra-excavated material must be restored to a density level compatible with surroundings to prevent excess deformation and/or joint rotation adjacent to the structure. In these areas, compact backfill to achieve the same soil density as specified for all pipe backfill but not less than required to achieve a soil modulus (M_{sb}) of at least 1,000 psi (6.9 MPa). The use of cement-stabilized backfills adjacent to large structures has been found to be effective in preventing excess deformation where diameters are larger than about 60 in. (1,500 mm). Other methods of accommodating the differential settlements may be acceptable.

Vertical risers. Provide support for vertical risers as commonly found at service connections, cleanouts, and drop manholes to preclude vertical or lateral movement. Prevent the direct transfer of thrust due to surface loads and settlement and ensure adequate support at points of connection to main lines.

Exposing pipe for making service line connections. When excavating for a service line connection, excavate material from above the top of the existing pipe before removing material from the sides of the pipe. When backfilling excavations of existing lines, the materials and construction methods used should restore the installation to its condition prior to excavation.

Pipe caps and plugs. Secure caps and plugs to the pipe to prevent movement and resulting leakage under test and service pressures. If lines are to be tested under pressure, any plugs and caps must be designed to safely carry the test pressure.

Parallel piping systems. Compact the soil between the pipes in the same manner as when compacting the soil between the pipe and the trench wall, taking special care to compact the soil in the haunches.

6.7.5 Thrust Blocks

Installation requirements related to thrust blocks are discussed in chapter 7.

6.8 FIELD MONITORING

Compliance with installation requirements for trench depth, grade, water conditions, foundation, embedment and backfill materials, joints, density of materials in place, and safety should be monitored to assure conformance with the contract documents.

Deflection. Monitor the deflection level in the pipe throughout the installation process for conformance to the requirements of the contract specifications and the manufacturer's recommendations. Conduct deflection measurement programs early in a project to verify that the construction procedures being used are adequate. The allowable deflection at the time of installation is the long-term allowable deflection reduced by the effects of deflection lag. If necessary, also consider the effects of vertical ovalling during compaction. Complete all deflection checks prior to conducting any pressure tests.

Pressure testing. Most pressure pipelines are tested after installation to detect leaks, installation flaws, damaged pipes, or other deficiencies. As a general rule, such tests should not be conducted using air pressure unless special precautions, not within the scope of this manual, are used. Additional recommendations for conducting pressure tests include:

- Required thrust restraints are properly installed (and sufficiently cured if applicable).

- Backfilling should be completed. Some sections of the line may be left uncovered provided suitable lateral and longitudinal restraint is provided.

- Pumps and valves are anchored.

- Assure test caps and endplugs are properly installed and restrained as necessary.

- Vent the pipeline while filling to allow all air to escape.

- Pressurize the line slowly to avoid pressure surges.

- In determining the test pressure, remember that the lowest point on the line will have the highest pressure. If the test pressure gauge is not installed at this location, determine the correct pressure by calculation.

- Assure that the test fluid temperature is stable during the test period (to avoid pressure changes due to thermal expansion or contraction that may be misinterpreted as leaks).

6.9 CONTRACT DOCUMENT RECOMMENDATIONS

The following guidelines may be included in contract documents for a specific project to cover installation requirements; ASTM D3839 provides similar guidelines. In either case, applications for a particular project may require that the engineer provide more specific requirements in several areas, including:

- maximum particle size if different from Sec. 6.5.2;
- restrictions on use of categories of embedment and backfill materials;
- specific gradations of embedment materials for resistance to migration;
- state-specific restrictions on leaving trenches open;
- restrictions on mode of dewatering and design of underdrains;
- requirements on minimum trench width;
- restrictions or details for support of trench walls;
- specific bedding and foundation requirements;
- specific restrictions on methods of compaction;
- minimum embedment density if different from these recommendations (specific density requirements for backfill [e.g., for pavement subgrade]);
- minimum cover requirements;
- detailed requirements for support of vertical risers, standpipes, and stacks to accommodate anticipated relative movements between pipe and appurtenances. Detailing to accommodate thermal movements, particularly at risers;
- detailed requirements for manhole connections;
- requirements on methods of testing compaction and leakage; and
- requirements on deflection and deflection measurements, including method and time of testing.

REFERENCE

American Association of State Highway and Transportation Officials. 1998. *AASHTO LRFD Bridge Design Specifications*. Second ed., with Interim Specifications through 2002. Washington, D.C.: American Association of State Highway and Transportation Officials.

AWWA MANUAL M45

Chapter 7

Buried Pipe Thrust Restraints

7.1 UNBALANCED THRUST FORCES

Unbalanced thrust forces occur in pressure pipelines at changes in direction (i.e., elbows, wyes, tees, etc.), at changes in cross-sectional area (i.e., reducers), or at pipeline terminations (i.e., bulkheads). These forces, if not adequately restrained, may cause pipeline movement resulting in separated joints and/or pipe damage. Thrust forces are: (1) hydrostatic thrust due to internal pressure of the pipeline and (2) hydrodynamic thrust due to changing momentum of flowing fluid. Since most pressure lines operate at relatively low velocities, the hydrodynamic force is very small and is usually ignored.

The equations in this chapter are presented with inch-pound units in the left-hand column and metric units in the right-hand column.

7.1.1 Hydrostatic Thrust

Typical examples of hydrostatic thrust are shown in Figure 7-1. The thrust in dead ends, tees, laterals, and reducers is a function of internal pressure P and cross-sectional area A at the pipe joint. The resultant thrust at a bend is also a function of the deflection angle Δ and is given by:

$$T = 2PA \sin (\Delta/2) \qquad T = 2{,}000 PA \sin (\Delta/2) \qquad (7\text{-}1)$$

Where:
T = hydrostatic thrust, lb
P = internal pressure, psi

Where:
T = hydrostatic thrust, N
P = internal pressure, kPa

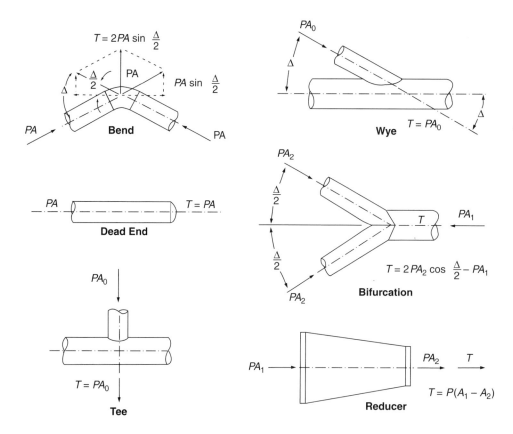

Figure 7-1   Thrust force definitions

A = cross-sectional area of pipe joint, in.^{2}
    = $(\pi/4) D_j^2$
D_j = joint diameter, in.
Δ = bend angle, degrees

A = cross-sectional area of pipe joint, m^2
    = $(\pi/4) (D_j/1{,}000)^2$
D_j = joint diameter, mm
Δ = bend angle, degrees

7.2   THRUST RESISTANCE

For buried pipelines, unbalanced horizontal thrust forces have two inherent sources of resistance: (1) frictional drag from dead weight of the pipe, earth cover, and contained fluid and (2) passive resistance of soil against the pipe or fitting in the direction of the thrust. If this resistance is not sufficient to resist the thrust, it must be supplemented by increasing the supporting area on the bearing side of the fitting with a thrust block; increasing the frictional drag of the line by "tying" adjacent pipe to the fitting; or otherwise anchoring the fitting to limit or prevent movement. Unbalanced uplift thrust at a vertical deflection is resisted by the dead weight of the fitting, earth cover, and contained fluid. If this type of resistance is not sufficient to resist the thrust, it must be supplemented by increasing the dead weight with a gravity-type thrust block; increasing the dead weight of the line by "tying" adjacent pipe to the fitting; or otherwise anchoring the fitting to limit or prevent movement.

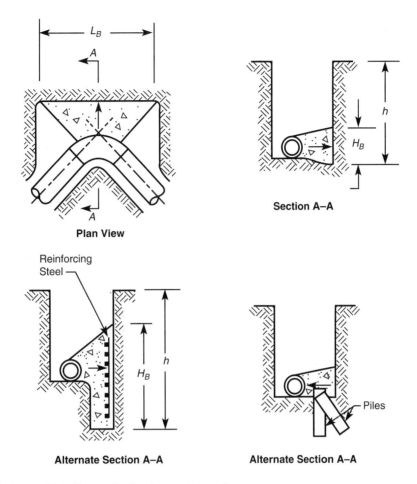

Figure 7-2 Typical thrust blocking of a horizontal bend

7.3 THRUST BLOCKS

Concrete thrust blocks increase the ability of fittings to resist movement by increasing the bearing area and the dead weight of the fitting. Typical thrust blocking of a horizontal bend (elbow) is shown in Figure 7-2.

7.3.1 Calculation of Size

Ignoring the dead weight of the thrust block, the block size can be calculated based on the bearing capacity of the soil:

Area of block = $(L_B)(H_B) = (T \times FS)/\sigma$ Area of block = $(L_B)(H_B) = (T \times FS)/1{,}000\,\sigma$ (7-2)

Where:
$(L_B)(H_B)$ = area of bearing surface of thrust block, ft^2
T = thrust force, lb
FS = design factor, 1.5
σ = bearing strength of soil, lb/ft^2

Where:
$(L_B)(H_B)$ = area of bearing surface of thrust block, m^2
T = thrust force, N
FS = design factor, 1.5
σ = bearing strength of soil, kPa

96   FIBERGLASS PIPE DESIGN

Table 7-1   Horizontal soil-bearing strengths

	Bearing Strength	
Soil	σ lb/ft^{2*}	kN/m^2
Muck	0	0
Soft clay	1,000	48
Silt	1,500	72
Sandy silt	3,000	144
Sand	4,000	192
Sandy clay	6,000	287
Hard clay	9,000	431

* Although the bearing strength values have been used successfully in the design of thrust blocks and are considered to be conservative, their accuracy is dependent on accurate soil identification and evaluation. The design engineer must select the proper bearing strength of a particular soil type.

Typical values for conservative horizontal bearing strengths of various soil types are listed in Table 7-1.

If it is impractical to design the block for the thrust force to pass through the geometric center of the soil-bearing area, the design should be evaluated for stability.

After calculating the concrete thrust block size, and reinforcement if necessary, based on the bearing capacity of soil, the shear resistance of the passive soil wedge behind the thrust block should be checked because it may govern the design. For a thrust block having its height H_B less than one-half the distance from the ground surface to base of block h, the design of the block is generally governed by the bearing capacity of the soil. However, if the height of the block H_B exceeds one-half h, the design of the block is generally governed by shear resistance of the soil wedge behind the thrust block. Determining the value of the bearing and shear resistance of the soil and thrust block reinforcement is beyond the scope of this manual. Consulting a qualified geotechnical engineer is recommended.

7.3.1.1 Typical configurations. Determining the bearing value σ is the key to "sizing" a thrust block. Values can vary from less than 1,000 lb/ft^2 (48 kN/m^2) for very soft soils to several tons per square foot (kN/m^2) for solid rock. Knowledge of local soil conditions is necessary for proper sizing of thrust blocks. Figure 7-2 shows several details for distributing thrust at a horizontal bend. Section A–A is the more common detail, but the other methods shown in the alternate sections may be necessary in weaker soils. Figure 7-3 illustrates typical thrust blocking of vertical bends. Design of the block for a bottom bend is the same as for a horizontal bend, but the block for a top bend must be sized to adequately resist the vertical component of thrust with dead weight of the block, bend, water in the bend, and overburden.

7.3.1.2 Proper construction is essential. Most thrust block failures can be attributed to improper construction. Even a correctly sized block can fail if it is not properly constructed. A block must be placed against undisturbed soil and the face of the block must be perpendicular to the direction of and centered on the line of action of the thrust. A surprising number of thrust blocks fail because of inadequate design or improper construction. Many people involved in construction and design do not realize the magnitude of the thrusts involved. As an example, a thrust block behind a 36-in. (900-mm), 90° bend operating at 100 psi (689 kPa) must resist a thrust force in excess of 150,000 lb (667 kN). Another factor frequently overlooked is that thrust increases in proportion to the square of pipe diameter. A 36-in. (900-mm) pipe produces

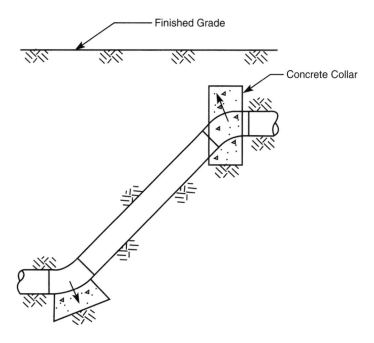

Figure 7-3 Typical profile of vertical bend thrust blocking

approximately four times the thrust produced by an 18-in. (450-mm) pipe operating at the same internal pressure.

7.3.2 Adjacent Excavation

Even a properly designed and constructed thrust block can fail if the soil behind the block is disturbed. Properly sized thrust blocks have been poured against undisturbed soil only to fail because another utility or an excavation immediately behind the block collapsed when the line was pressurized. If the risk of future nearby excavation is high, the use of restrained (tied) joints may be appropriate.

7.4 JOINTS WITH SMALL DEFLECTIONS

The thrust at pipe joints installed with angular deflection is usually so small that supplemental restraint is not required.

7.4.1 Small Horizontal Deflections

Thrust T at horizontal deflected joints is resisted by friction on the top and bottom of the pipe, as shown in Figure 7-4. Additional restraint is not required when:

$$T \leq f L_p (W_p + W_w + 2W_e) \qquad T \leq f L_p (W_p + W_w + 2W_e) \qquad (7\text{-}3)$$

Where:
$T = 2PA \sin(\theta/2)$, lb
θ = angle of deflected joint, degrees
f = coefficient of friction

Where:
$T = 2{,}000 PA \sin(\theta/2)$, N
θ = angle of deflected joint, degrees
f = coefficient of friction

98   FIBERGLASS PIPE DESIGN

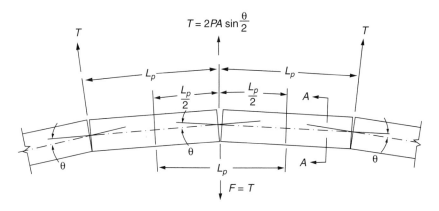

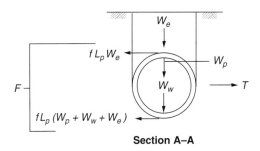

Figure 7-4   Restraint of thrust at deflected joints on long-radius horizontal curves

L_p = length of pipe, ft L_p = length of pipe, m
W_p = weight of pipe, lb/lin ft W_p = weight of pipe, N/m
W_w = weight of fluid in pipe, lb/lin ft W_w = weight of fluid in pipe, N/m
W_e = earth cover load, lb/lin ft W_e = earth cover load, N/m

The passive soil resistance of the trench backfill against the pipe is ignored in the previous analysis. Depending on the installation and field conditions, the passive soil resistance of the backfill may be included to resist thrust.

The selection of a value for the coefficient of friction f is dependent on the type of soil and the roughness of the pipe exterior. Design values for the coefficient of friction generally vary from 0.25 to 0.50.

Determination of earth cover load should be based on a backfill density and height of cover consistent with what can be expected when the line is pressurized. Values of soil density vary from 90 lb/ft^3 to 130 lb/ft^3 (14 kN/m^3 to 20 kN/m^3), depending on the degree of compaction. W_e may be conservatively determined using the Marston equation for loads imparted to a flexible pipe, as follows:

$$W_e = (C_d)(W)(B_d)(B_c) \qquad\qquad W_e = (C_d)(W)(B_d)(B_c) \qquad (7\text{-}4)$$

Where: Where:
    W_e = earth load, lb/lin ft of pipe length W_e = earth load, N/m of pipe length

C_d = coefficient based on soil and the ratio of H and B_d (see Figure 7-5)
H = depth of fill to pipe top, ft
W = unit weight of soil, lb/ft³
B_d = trench width at pipe top, ft
B_c = OD of pipe, ft

C_d = coefficient based soil and the ratio of H and B_d (see Figure 7-5)
H = depth of fill to pipe top, m
W = unit weight of soil, N/m³
B_d = trench width at pipe top, m
B_c = OD of pipe, m

7.4.2 Small Vertical Deflections With Joints Free to Rotate

Uplift thrust at deflected joints on long-radius vertical curves is resisted by the combined dead weight W_t, as shown in Figure 7-6. Additional restraint is not required when:

$$T \le L_p (W_p + W_w + W_e) \cos (\varphi - \theta/2) \qquad T \le L_p (W_p + W_w + W_e) \cos (\varphi - \theta/2) \qquad (7\text{-}5)$$

Where:
$T = 2PA \sin (\theta/2)$, lb
L_p = length of standard or beveled pipe, ft
φ = slope angle, degrees
θ = deflection angle of joint, degrees

Where:
$T = 2{,}000 PA \sin (\theta/2)$, N
L_p = length of standard or beveled pipe, m
φ = slope angle, degrees
θ = deflection angle of joint, degrees

7.5 RESTRAINED (TIED) JOINTS

Unbalanced thrust forces at fittings or deflected joints may be resisted by using restrained joint(s) across the deflected joint or by tying adjacent pipes to the fitting. This method fastens a number of pipes on each side of the fitting to increase the frictional drag of the connected pipe to resist the fitting thrust. Since thrust diminishes from a maximum value at a fitting to zero at distance L from the fitting, requirements for longitudinal strength to resist thrust can be calculated for the pipe length immediately adjacent to the fitting and prorated on a straight-line basis for the remainder of the pipe within the tied distance L. Frictional resistance on the tied pipe acts in the opposite direction of resultant thrust T. Section A–A in Figure 7-4 shows the external vertical forces acting on a buried pipe with horizontal thrust and the corresponding frictional resistance. Uplift thrust restraint provided by gravity-type thrust blocks, shown for the top bend in Figure 7-3, may also be provided by the alternate method of increasing the dead weight of the line by tying adjacent pipe to the vertical bend. Section A–A in Figure 7-6 shows the vertical forces acting on a buried vertical (uplift) bend used in determining the thrust resistance by dead weight.

As previously stated, both of these analyses ignore the passive soil resistance of the backfill against the pipe. Depending on the installation and field conditions, the passive soil resistance of the backfill may be included to resist thrust.

7.5.1 Horizontal Bends and Bulkheads

As illustrated in Figure 7-7, the frictional resistance F needed along each leg of a horizontal bend is $PA\sin(\Delta/2)$. Frictional resistance per linear foot of pipe against soil is equal to:

100 FIBERGLASS PIPE DESIGN

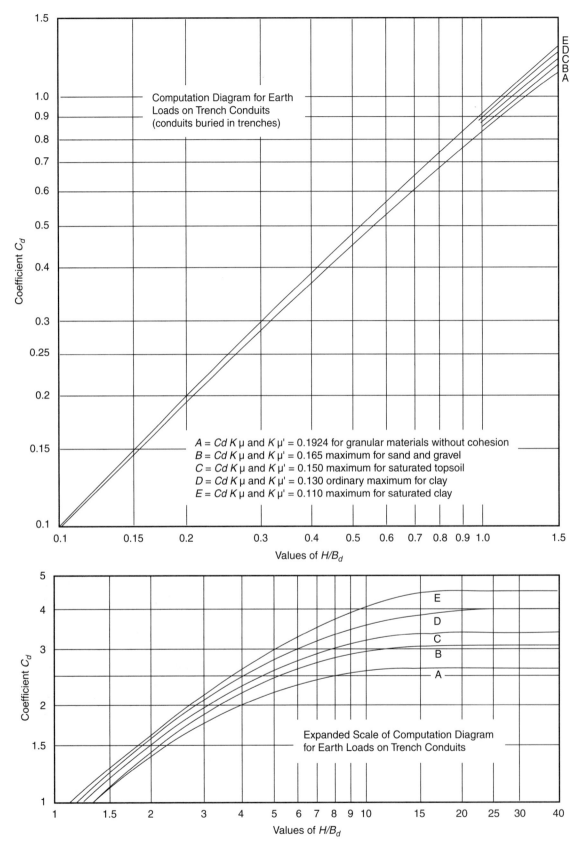

Figure 7-5 Computation diagram for earth loads on trench conduits

BURIED PIPE THRUST RESTRAINTS 101

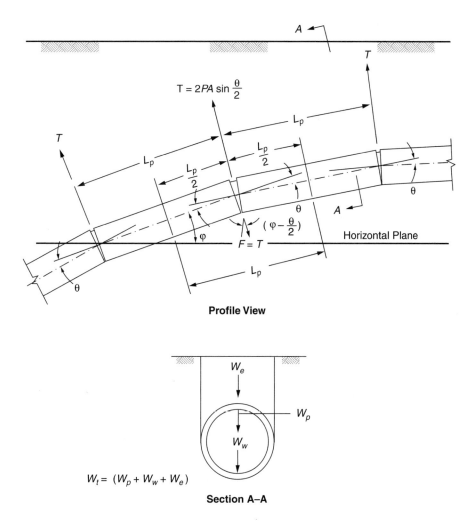

Figure 7-6 Restraint of uplift thrust at deflected joints on long-radius vertical curves

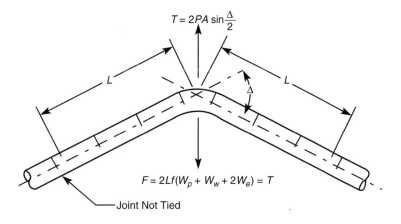

Figure 7-7 Thrust restraint with tied joints at bends

$$F = f(2W_e + W_p + W_w) \qquad F = f(2W_e + W_p + W_w) \qquad (7\text{-}6)$$

Where: Where:

f = coefficient of friction between pipe and soil

F = frictional resistance, lb/ft

f = coefficient of friction between pipe and soil

F = frictional resistance, N/m

Therefore, the length of pipe L to be tied to each leg of a bend is calculated as:

$$L_{\text{bend}} = \frac{PA\sin(\Delta/2)}{f(2W_e + W_p + W_w)} \qquad L_{\text{bend}} = \frac{1{,}000 PA\sin(\Delta/2)}{f(2W_e + W_p + W_w)} \qquad (7\text{-}7)$$

Where: Where:

L_{bend} = length of pipe tied to each bend leg, ft

L_{bend} = length of pipe tied to each bend leg, m

The length of pipe to be tied to a bulkhead or tee leg is:

$$L_{\text{bulk}} = \frac{PA}{f(2W_e + W_p + W_w)} \qquad L_{\text{bulk}} = \frac{PA}{f(2W_e + W_p + W_w)} \qquad (7\text{-}8)$$

Where: Where:

L_{bulk} = length of pipe tied to bulkhead or tee leg, ft, with all other variables as defined previously

L_{bulk} = length of pipe tied to bulkhead or tee leg, m, with all other variables as defined previously

7.5.2 Vertical (Uplift) Bends

As illustrated in Figure 7-8, the dead weight resistance needed along each leg of a vertical bend is $2PA\sin(\Delta/2)$. Dead weight resistance per linear foot of pipe in a direction opposite to thrust is:

$$D_{wr} = (W_e + W_p + W_w)\cos(\varphi - \Delta/2) \qquad D_{wr} = (W_e + W_p + W_w)\cos(\varphi - \Delta/2) \qquad (7\text{-}9)$$

Where: Where:

D_{wr} = dead weight resistance, lb/ft

φ = slope angle, degrees (See Figure 7-8)

Δ = bend angle, degrees (See Figure 7-8)

D_{wr} = dead weight resistance, N/m

φ = slope angle, degrees (See Figure 7-8)

Δ = bend angle, degrees (See Figure 7-8)

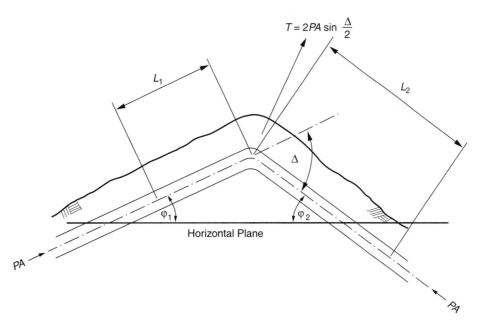

Figure 7-8 Length of tied pipe on each leg of vertical (uplift) bend

Length of pipe L to be tied to leg of a vertical (uplift) bend is calculated as:

$$L_t = \frac{PA[\sin(\Delta/2)]}{(W_e + W_p + W_w)\cos[\varphi - (\Delta/2)]} \qquad L_t = \frac{1{,}000\,PA[\sin(\Delta/2)]}{(W_e + W_p + W_w)\cos[\varphi - (\Delta/2)]} \quad (7\text{-}10)$$

Where:
  L_t = length of pipe to be tied to one of the legs of a vertical (uplift) bend, ft (See Figure 7-8 and note that angles φ_1 and φ_2 may be equal or different.)

Where:
  L_t = length of pipe to be tied to one of the legs of a vertical (uplift) bend, m (See Figure 7-8 and note that angles φ_1 and φ_2 may be equal or different.)

Vertical downward bends are resisted by bearing of the trench against the bottom of the pipe. Properly bedded pipe should not have to be investigated for this condition.

7.5.3 Transmission of Thrust Force Through Pipe

In addition to calculating pipe length to be tied to a fitting, engineers must be sure that tied pipe lengths have sufficient strength in the longitudinal direction to transmit thrust forces. The maximum thrust force for which the pipe adjacent to a bend must be designed is equal to:

$$F_y = 0.001(5.43\Delta + 0.063\Delta^2)\,PA \qquad F_y = (5.43\Delta + 0.063\Delta^2)\,PA \qquad (7\text{-}11)$$

Where:
  F_y = maximum axial thrust force for which the pipe adjacent to a bend must be designed, lb

Where:
  F_y = maximum axial thrust force for which the pipe adjacent to a bend must be designed, N

Note:
  For $\Delta = 90°$; $F_y = PA$

Note:
  For $\Delta = 90°$; $F_y = PA$

This page intentionally blank.

AWWA MANUAL M45

Chapter **8**

Aboveground Pipe Design and Installation

8.1 INTRODUCTION

This chapter addresses the design and installation of fiberglass pipeline systems in aboveground applications for pipe diameters of 16 in. (400 mm) and smaller and only for pipe lines that have restrained joints. Different design provisions and supporting methods may be applicable for specific project requirements, larger diameters, or a particular piping product. Consult with the manufacturer and the piping engineer for appropriate design considerations.

The equations in this chapter are presented with inch-pound units in the left-hand column and metric units in the right-hand column.

8.2 THERMAL EXPANSION AND CONTRACTION

Fiberglass pipe may have a different expansion rate in the hoop and axial directions. For example, a filament-wound pipe with a 55° winding angle has about the same thermal expansion as steel in the hoop direction and about twice the thermal expansion as steel in the axial direction. The total expansion or contraction for a pipe system is determined using the following equation:

$$L_c = (12)(C_t)(L)(T_c) \qquad\qquad L_c = (1{,}000)(C_t)(L)(T_c) \qquad (8\text{-}1)$$

Where:

L_c = length change, in.

C_t = coefficient of axial thermal expansion in./in./°F

Where:

L_c = length change, mm

C_t = coefficient of axial thermal expansion mm/mm/°C

L = length between anchors, ft

T_c = temperature change, °F
(maximum operating temperature minus installation temperature for expansion and installation temperature minus minimum operating temperature for contraction)

L = length between anchors, m

T_c = temperature change, °C
(maximum operating temperature minus installation temperature for expansion and installation temperature minus minimum operating temperature for contraction)

8.3 THERMAL EXPANSION DESIGN

In the design of aboveground pipelines, the supports and guides for the pipe become important considerations because of thermal expansion.

In addition to pressure resistance and life limitations, the effects of thermal expansion and contraction should be considered. A number of methods accommodate the length changes associated with thermal expansion and contraction. The four most commonly used methods include:

- anchoring and guiding,
- direction changes,
- expansion loops, and
- mechanical expansion joints.

Guides, expansion loops, and mechanical expansion joints are installed in straight lines and are anchored at each end. Experience has shown that direction changes are the least expensive method of accommodating thermal expansion. Guide spacing is the next most economical method, followed by mechanical expansion joints and expansion loops.

For small temperature changes and piping systems that consist of short run lengths, it is usually unnecessary to make special provisions for thermal expansion. However, any system should have the capability of accommodating length changes.

Experience has shown that aboveground piping systems need anchors at approximately 300-ft (90-m) intervals. (NOTE: This value may vary for larger pipe sizes.) These anchors limit pipe movement caused by vibrations and transient loading conditions. Anchors should fasten all transition points within the system. Transition points are places where pipe diameter, material, elevation, or direction changes or where manufacturer changes. Anchors at transition points limit the transfer of thermal end loads from line section to line section.

8.3.1 Thermal End Loads

The axial modulus of elasticity of fiberglass pipe can vary from approximately 1.5% to 10% of steel. The low modulus results in lower end loads that require restraining equipment less strong than that used for metallic pipelines. Internal pressures in the piping system can result in some length change. Experience has shown that this elongation is often insignificant and may not need to be considered in the design.

The equation for calculating the thermal end load is:

$$E_L = (C_t)(E)(A)(T_c) \qquad\qquad E_L = (C_t)(E)(A)(T_c) \times 10^9 \qquad (8\text{-}2)$$

Where:

E_L = thermal end load, lb
E = axial modulus, psi
A = cross-sectional area of pipe wall, in.^{2}
   = $\pi/4(OD^2 - ID^2)$
   OD = pipe outside diameter, in.
   ID = pipe inside diameter, in.

Where:

E_L = thermal end load, N
E = axial modulus, GPa
A = cross-sectional area of pipe wall, m^2
   = $\pi/4(OD^2 - ID^2) \times 10^{-6}$
   OD = pipe outside diameter, mm
   ID = pipe inside diameter, mm

When pipe between anchors expands, the pipe undergoes compression. When contraction occurs, the pipe experiences tension.

8.3.2 Spacing Design—Anchoring and Guiding

Installing anchors at all directional and elevation changes serves to divide the system into straight runs. With anchors installed, guides are an economical method for dealing with expansion. The relatively low modulus of fiberglass pipe allows it to absorb the thermal stresses as compressive stresses in the pipe wall. Compressive stresses from expansion may result in buckling, unless the pipe is constrained at close intervals to prevent columnar instability.

The equation to calculate maximum allowable guide spacing interval is:

$$L_G = \sqrt{\frac{(\pi^2)(E_b)(I)}{144(C_t)(A)(T_c)(E_c)}} \qquad L_G = \sqrt{\frac{(\pi^2)(E_b)(I)}{(C_t)(A)(T_c)(E_c)}} \qquad (8\text{-}3)$$

Where:

L_G = maximum guide (support) spacing, ft
E_b = axial bending modulus, psi
I = moment of inertia, in.^{4}
   = $\dfrac{\pi(OD^4 - ID^4)}{64}$
E_c = axial compressive modulus, psi

Where:

L_G = maximum guide (support) spacing, m
E_b = axial bending modulus, GPa
I = moment of inertia, mm^4
   = $\dfrac{\pi(OD^4 - ID^4) \times 10^{-12}}{64}$
E_c = axial compressive modulus, GPa

Because the bending and compressive moduli are obtained from experimental data, the ratio E_b/E_c, using data representative of the minimum and maximum operating temperatures, should be calculated. The lower value of the two calculations will satisfy the interest of conservative design.

Compare guide intervals with the intervals for supports, then adjust guide spacing for a better match with support spacing. For example, adjust intervals so a guide replaces every second or third support. Remember, all guides act as modified supports and must meet the minimum requirements for supports, anchors, and guides, as prescribed in other sections of this chapter.

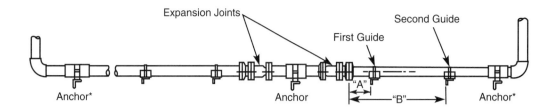

*Anchor Load = $\frac{\pi}{4}$ (ID)^2 × Internal Pressure

NOTE: A = 4 diameters; B = 14 diameters

Reprinted with permission from Fiberglass Pipe Handbook, *Fiberglass Pipe Institute, New York, N.Y.*

Figure 8-1 Typical expansion joint installation

8.3.3 Expansion Joint Design

Expansion joints may be used to absorb thermal expansion in long, straight pipe runs. Various types of expansion joints are available and suitable for use with fiberglass piping systems. Because the forces developed during a temperature change are relatively low compared with metallic systems, it is essential to specify an expansion joint that activates with low force. Remember that fiberglass pipe will expand more than most metallic systems. The required movement per expansion joint and the number of expansion joints may be greater for fiberglass systems.

The allowable activation force for expansion joints depends on both the thermal forces developed in the pipe and the support or guide spacing. Guide spacing at the entry of an expansion joint is typically 4 pipe diameters (first guide) and 14 pipe diameters (second guide) from the inlet of the expansion joint (Figure 8-1). These guides and locations give proper alignment. The spacing of the remaining supports should remain within the maximum calculated interval.

The equation for calculating the allowable activation force is:

$$P_{cr} = \frac{\pi^2 (E_c)(I)(S_f)}{L_G^2} \qquad P_{cr} = \frac{\pi^2 (E_c)(I)(S_f) \times 10^9}{L_G^2} \qquad (8\text{-}4)$$

Where:

P_{cr} = critical buckling force of pipe, lb

S_f = material variation safety factor; 0.9 recommended

Where:

P_{cr} = critical buckling force of pipe, N

S_f = material variation safety factor; 0.9 recommended

The pressure thrust must also be considered. Pressure thrust is the design pressure times the area of the expansion joint.

In all applications, the activation force of the expansion joint must not exceed the thermal end loads developed by the pipe. The cost and limited motion capability of expansion joints makes them impractical to use in many applications. In these cases, loops, guide spacing, or short lengths of flexible hose can handle thermal expansion. The expansion joint needs an anchor on both sides for proper operation.

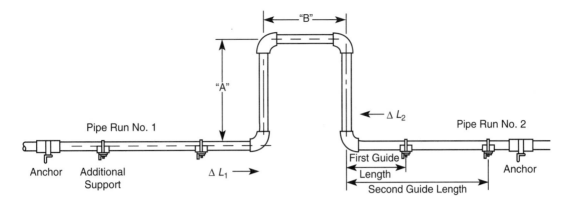

Reprinted with permission from Fiberglass Pipe Handbook, *Fiberglass Pipe Institute, New York, N.Y.*

Figure 8-2 Expansion loop dimensions

8.3.4 Expansion Loop Design

Expansion loops flex to accommodate changes in length (Figure 8-2). This design method is used to calculate the stress developed in a cantilevered beam with a concentrated load at the free end and ignores flexibility of the loop leg, the leg parallel to the line.

Two guides on both sides of each expansion loop ensure proper alignment. The recommended guide spacing is 4 (first guide) and 14 (second guide) nominal pipe diameters. Additional guides or supports should be located so the maximum spacing interval is not exceeded.

To design an expansion loop, use the following equation:

$$L_A = \sqrt{\frac{(K)(L_c)(E_b)(\text{OD})}{(144)(\sigma_b)}} \qquad L_A = \sqrt{\frac{(K)(L_c)(E_b)(\text{OD})}{(1{,}000)(\sigma_b)}} \qquad (8\text{-}5)$$

Where:

L_A = length of the "A" leg, ft

K = cantilevered beam constant

= 0.75 for nonguided beam

= 3.0 for guided beam

σ_b = design allowable bending stress, psi (A minimum safety factor of 8 on ultimate strength is recommended.)

Where:

L_A = length of the "A" leg, m

K = cantilevered beam constant

= 0.75 for nonguided beam

= 3.0 for guided beam

σ_b = design allowable bending stress, MPa (A minimum safety factor of 8 on ultimate strength is recommended.)

If the maximum allowable bending stress of the fittings is greater than the maximum for the pipe, the bending moment of the fitting does not need to be considered. In other cases, the fitting manufacturer will provide allowable bending moments for the fittings. These values are used in Eq 8-6 to determine the "A" leg length. The results are compared and the larger value is used. Pipelines with heavy-wall pipe and relatively thin-wall fittings are most likely to require verification of the L_A dimension.

$$L_A = \sqrt{\frac{12(L_c)(E_b)(I)}{144(M)}} \qquad L_A = \sqrt{\frac{(L_c)(E_b)(I) \times 10^6}{M}} \qquad (8\text{-}6)$$

Where:
L_c = maximum length change (see Eq 8-1), in.
M = allowable elbow bending moment, lb-in.

Where:
L_c = maximum length change (see Eq 8-1), mm
M = allowable elbow bending moment, N-m

In some cases, the manufacturer may require anchors at all fittings. For example, mitered fittings and/or large-diameter fittings may have allowable bending stresses below that of the pipe. In these cases, thermal expansion procedures may be limited to the use of anchors and guides or expansion joints if the bending moment is not available.

8.3.5 Direction Changes

In some installations, system directional changes can perform the same function as expansion loops. Directional changes that involve some types of fittings, such as saddles, should not be used to absorb expansion or contraction. The bending stresses may cause fitting failure. Stress in the pipe at a given directional change depends on the total change in length and the distance to the first secure hanger or guide past the directional change. In other words, the required flexible leg length is based on the maximum change in length.

Recommended support or guide spacing cannot be disregarded. However, flexible or movable supports, such as strap hangers, can provide support while allowing the pipe to move and absorb the changes in length. Supports must prevent lateral movement or pipe buckling.

Where large thermal movements are expected, a short length of flexible hose installed at a change in direction will absorb some of the line movement. This method of handling thermal expansion is usually the most economical means of compensating for large displacements when the guide spacing method cannot be used. Hose manufacturers provide specifications giving the minimum bend radius, chemical compatibility, temperature, and pressure rating of a particular flexible hose.

The equation for calculating the length of the flexible pipe leg (i.e., the distance to the first restraining support or guide) is:

$$L_{sh} = \sqrt{\frac{1.5(L_c)(E_b)(\text{OD})}{(144)(\sigma_b)}} \qquad L_{sh} = \sqrt{\frac{1.5(L_c)(E_b)(\text{OD})}{(1{,}000)(\sigma_b)}} \qquad (8\text{-}7)$$

Where:
L_{sh} = length from direction change to the first secure hanger, ft

Where:
L_{sh} = length from direction change to the first secure hanger, m

This type of analysis usually neglects torsional stresses. Allowable bending stress is much lower than the allowable torsional stress. Therefore, bending of the pipe leg, as

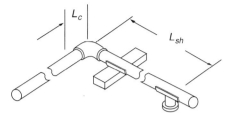

Figure 8-3 Directional change

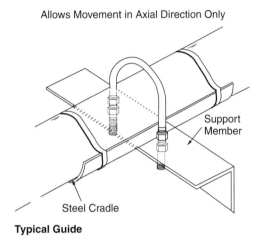

Typical Guide

Reprinted with permission from Fiberglass Pipe Handbook, *Fiberglass Pipe Institute, New York, N.Y.*

Figure 8-4 Guide support

shown in Figure 8-3, will typically absorb pipe movement. However, the unanchored leg must have a free length equal to or greater than L_{sh}, as calculated from Eq 8-7.

8.4 SUPPORTS, ANCHORS, AND GUIDES

8.4.1 Guides

The guiding mechanism must be loose to allow free axial movement of the pipe. However, the guides must be attached rigidly to the supporting structure so that the pipe moves only in the axial direction (Figure 8-4).

All guides act as supports and must meet the minimum requirements for supports. Refer to Sec. 8.4.6 if thermal cycles are frequent.

8.4.2 Anchors

An anchor must restrain the movement of the pipe against all applied forces. Pipe anchors divide a pipe system into sections. They attach to structural material capable of withstanding the applied forces. In some cases, pumps, tanks, and other similar equipment function as anchors. However, most installations require additional anchors where pipe sizes change and fiberglass pipe joins another material or a product from another manufacturer. Additional anchors usually occur at valve locations,

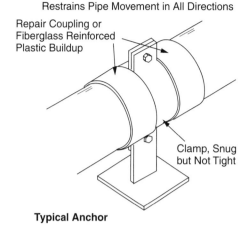

Reprinted with permission from Fiberglass Pipe Handbook, *Fiberglass Pipe Institute, New York, N.Y.*

Figure 8-5   Anchor support

changes in direction of piping runs, and major branch connections. Saddles and laterals are particularly sensitive to bending stresses. To minimize stresses on saddles and laterals, anchor the pipe on either side of the saddle or anchor the side run.

Figure 8-5 shows a typical anchor. Operating experience with piping systems indicates that it is a good practice to anchor long, straight runs of aboveground piping at approximately 300-ft (90-m) intervals. These anchors prevent pipe movement due to vibration or water hammer.

One anchoring method features a clamp placed between anchor sleeves or a set of anchor sleeves and a fitting. The sleeves bonded on the pipe prevent movement in either direction. Sleeve thickness must equal or exceed the clamp thickness. To achieve this, it often is necessary to bond two sleeves on each side of the clamp. Anchor sleeves are usually 1 pipe diameter in length and cover 180° of circumference. Anchors act as supports and guides and must meet minimum requirements for supports.

8.4.3   Supports

To prevent excessive pipe deflection due to the pipe and fluid weight, support horizontal pipe (see Figure 8-6) at intervals determined by one of the following methods.

8.4.3.1   Type I.   Pipe analyzed as simply supported single spans (two supports per span length) with the run attached to a fitting at one end, or any other section of less than three span lengths. Beam analysis for other types of spans, such as a section adjacent to an anchor, is sometimes used to obtain a more accurate span length. However, the following equation is more conservative:

$$L_s = \sqrt[4]{\frac{(d_m)(E_b)(I)}{0.013(W)}} \qquad L_s = \sqrt[4]{\frac{(d_m)(E_b)(I) \times 10^9}{13(W)}} \qquad (8\text{-}8)$$

Where:

L_s = unsupported span, in.

Where:

L_s = unsupported span, m

ABOVEGROUND PIPE DESIGN AND INSTALLATION 113

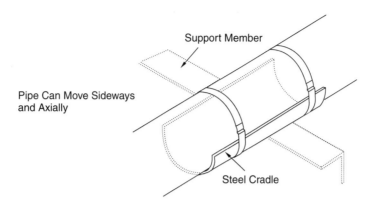

Reprinted with permission from Fiberglass Pipe Handbook, *Fiberglass Pipe Institute, New York, N.Y.*

Figure 8-6 Typical support

d_m = allowable midpoint deflection (typical value for fiberglass pipe is 0.5 in.), in.

$W = P_w + W_f$, lb/in. of length

  P_w = pipe weight, lb/in. of length

  W_f = fluid weight, lb/in. of length

  $= \rho V_p/12$

  ρ = fluid density, lb/ft^3

  V_p = pipe volume per foot of length, ft^3/ft

d_m = allowable midpoint deflection (typical value for fiberglass pipe is 12.5 mm), mm

$W = P_w + W_f$, N/m of length

  P_w = pipe weight, N/m of length

  W_f = fluid weight, N/m of length

  $= \rho V_p/12$

  ρ = fluid density, N/m^3

  V_p = pipe volume per meter of length, m^3/m

When mid-span deflection d_m exceeds 0.5 in. (13 mm), check with the pipe manufacturer for other considerations, such as the allowable bending stress or bearing stress. When the mid-span deflection is limited to 0.5 in. (13 mm), the bending stress on the pipeline is typically below the allowable bending stress for the pipe. For installations that result in more than 0.5 in. (13 mm) of mid-span deflection, the 8:1 safety factor on bending stress has proven to be sufficient to compensate for the combination of bending stress and the longitudinal stresses resulting from internal pressure.

In fact, cyclic bending tests have shown that the stresses are not additive as expected and that the 8:1 safety factor is conservative. Cyclic bending tests consist of cyclic pressure testing of pipe bent to stress levels at or above the design bending stress.

For low stiffness pipe with a relatively thin wall, the local bearing pressure at supports is often significant. Supports for this application usually require 180° contact and follow a conservative design-allowable bearing pressure (45 psi [310 kPa]) compared with the typically permitted 85 psi (586 kPa) used for smaller diameter, higher stiffness pipe. Because pipe design differs among manufacturers, follow the supplier's recommendations for the product and system.

8.4.3.2 Type II. Pipe analyzed as a continuous beam—three spans—all loaded.

$$L_s = \sqrt[4]{\frac{(d_m)(E_b)(I)}{(0.0069)(W)}} \qquad L_s = \sqrt[4]{\frac{(d_m)(E_b)(I) \times 10^9}{6.9(W)}} \qquad (8\text{-}9)$$

8.4.3.3 Type III. Pipe analyzed as a continuous beam—four spans—all spans loaded.

$$L_s = \sqrt[4]{\frac{(d_m)(E_b)(I)}{(0.0065)(W)}} \qquad L_s = \sqrt[4]{\frac{(d_m)(E_b)(I) \times 10^9}{6.5(W)}} \qquad (8\text{-}10)$$

8.4.3.4 Type IV. Pipe analyzed as a beam fixed at both ends—uniformly distributed loads.

$$L_s = \sqrt[4]{\frac{(d_m)(E_b)(I)}{(0.0026)(W)}} \qquad L_s = \sqrt[4]{\frac{(d_m)(E_b)(I) \times 10^9}{2.6(W)}} \qquad (8\text{-}11)$$

Supports must also meet the minimum requirements for supports described in Sec. 8.4.4 through Sec. 8.4.9.

NOTE: In cases where the wall thickness to diameter ratio is low, the possibility of buckling failures at the supports is a concern. This may require the use of empirical equations and special bearing stress calculations that were determined or verified by testing.

Six basic rules control design and positioning for supports, anchors, and guides. These are described in the following paragraphs.

8.4.4 Rule 1: Avoid Point Loads

Use curved supports fitted to contact the bottom 120° of the pipe and that have a maximum bearing stress of 85 psi (586 kPa). Do not allow unprotected pipe to press against roller supports, flat supports, such as angle iron or I-beams, or U-bolts. Do not allow pipe to bear against ridges or points on support surfaces. Use metal or fiberglass sleeves to protect pipe if these conditions exist.

8.4.5 Rule 2: Meet Minimum Support Dimensions

Standard pipe supports designed for steel pipe can support fiberglass pipe if the minimum support widths provided in Table 8-1 are met. Supports failing to meet the minimum must be augmented with a protective sleeve of split fiberglass pipe or metal. In all cases, the support must be wide enough that the bearing stress does not exceed 85 psi (586 kPa).

Sleeves augmenting supports must be bonded in place using adhesives stable at the system's maximum operating temperature.

Prepare all pipe and sleeve bonding surfaces by sanding the contacting surfaces.

8.4.6 Rule 3: Protect Against External Abrasion

If vibrations or pulsations are possible, protect contacting surfaces from wear (Figure 8-7). When frequent thermal cycles, vibrations, or pulsating loadings affect the pipe, all contact points must be protected. This is typically accomplished by bonding a wear saddle made of fiberglass, steel, or one half of a section of the same pipe to the wall.

Table 8-1 Minimum support width for 120° contact supports

Pipe Size		Minimum Support Width	
in.	*mm*	*in.*	*mm*
1	25	0.88	22.4
1.5	40	0.88	22.4
2	50	0.88	22.4
3	80	1.25	31.8
4	100	1.25	31.8
6	150	1.50	38.1
8	200	1.75	44.5
10	250	1.75	44.5
12	300	2.00	50.8
14	350	2.00	50.8
16	400	2.50	63.5

NOTE: Table is based on maximum liquid specific gravity of 1.25.

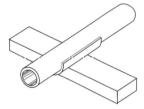

Reprinted with permission from Fiberglass Pipe Handbook, *Fiberglass Pipe Institute, New York, N.Y.*

Figure 8-7 Fiberglass wear protection cradle

8.4.7 Rule 4: Support Heavy Equipment Independently

Valves and other heavy equipment must be supported independently in both horizontal and vertical directions (Figure 8-8).

8.4.8 Rule 5: Avoid Excessive Bending

When laying lines directly on the surface, take care to ensure there are no excessive bends that would impose undue stress on the pipe.

8.4.9 Rule 6: Avoid Excessive Loading in Vertical Runs

Support vertical pipe runs as shown in Figure 8-9. The preferred method is to design for "pipe in compression." If the "pipe in tension" method cannot be avoided, take care to limit the tensile loadings below the recommended maximum tensile rating of the pipe. Install guide collars using the same spacing intervals used for horizontal lines (Figure 8-5).

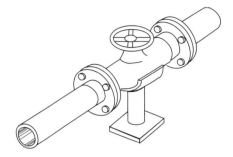

Reprinted with permission from Fiberglass Pipe Handbook, *Fiberglass Pipe Institute, New York, N.Y.*

Figure 8-8 Steel wear protection cradle

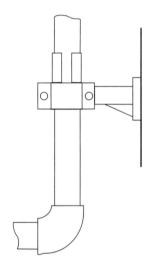

Reprinted with permission from Fiberglass Pipe Handbook, *Fiberglass Pipe Institute, New York, N.Y.*

Figure 8-9 Vertical support

8.5 BENDING

The minimum bending radius for fiberglass pipe usually is determined using a design stress that is one eighth of the ultimate short-term bending stress. Certain fittings, such as saddles and laterals, may be more susceptible to bending failure than other types. Consult the manufacturer for recommendations and limitations. The equation for calculating the minimum bending radius is:

$$R_m = \frac{(E_b)(\text{OD})}{24\sigma_b} \qquad R_m = \frac{(E_b)(\text{OD})}{2\sigma_b} \qquad (8\text{-}12)$$

Where:
  R_m = minimum allowable bending radius, ft

Where:
  R_m = minimum allowable bending radius, m

Because material properties vary with temperature, the allowable minimum bending radius will also vary.

8.6 THERMAL CONDUCTIVITY

The thermal conductivity of fiberglass pipe wall is approximately 1% that of steel. However, in most heat transfer situations, the heat loss or gain for pipe is controlled by the resistance to heat flow into the surrounding media (i.e., air or soil) rather than the thermal conductivity of the pipe. This reduces the insulating effect of a relatively thin fiberglass pipe wall. For this reason, thermal insulation tables for steel pipe can be used to size the insulation for most fiberglass pipelines.

The coefficient of thermal conductivity varies for different types of fiberglass pipe. A typical value for an epoxy resin pipe is 2.5–3.0 Btu/(hr) (ft^2) (°F)/in. (0.36–0.43 W/m-K). A typical value for polyester or vinyl ester resin pipe is 1.0–1.5 Btu/(hr) (ft^2) (°F)/in. (0.14–0.22 W/m-K).

8.7 HEAT TRACING

Both steam tracing and electrical heating tapes are acceptable techniques for heating fiberglass pipe. When using either method, three criteria govern the maximum element temperature:

1. The average wall temperature must not exceed the temperature rating of the pipe.
2. The maximum tracing temperature must not be more than 100°F (38°C) above the maximum rated temperature of the pipe.
3. The maximum recommended chemical resistance temperature of the pipe must not be exceeded at the inside wall of the pipe.

For criteria 1, the following equation is applicable:

$$A_t = (T_i + T_t) / 2 \qquad\qquad A_t = (T_i + T_t) / 2 \qquad (8\text{-}13)$$

Where:
- A_t = average wall temperature, °F
- T_i = inside wall temperature, °F
- T_t = heat tracing temperature, °F

Where:
- A_t = average wall temperature, °C
- T_i = inside wall temperature, °C
- T_t = heat tracing temperature, °C

For criteria 2, the following equation is used:

$$T_t = T_R + 100°F \qquad\qquad T_t = T_R + 56°C \qquad (8\text{-}14)$$

Where:
- T_R = maximum rated temperature of pipe, °F

Where:
- T_R = maximum rated temperature of pipe, °C

(NOTE: The 100°F is a Δ temp, which must be converted to metric by 0.56°C/°F.)

The maximum tracing element temperature is the lesser of the values calculated using Eq 8-13 and Eq 8-14.

The maximum tracing element temperature using this methodology applies only to applications involving flowing, nonstagnant, fluid conditions. For stagnant conditions, the maximum allowable trace element is the chemical resistance temperature of the pipe.

For criteria 3, it is necessary to check the manufacturer's published data to determine the maximum recommended chemical resistance of the pipe for this application. This value is compared with the inside wall temperature T_i. The published value must be greater than T_i.

8.8 CHARACTERISTICS AND PROPERTIES

The characteristics and properties for fiberglass pipe are different from those typically used for metallic pipes.

8.8.1 Design Pressure or Stress

Design stresses for pipe internal pressure are based on ASTM D2992. The internal operating pressure for fittings is generally based on one fourth of the ultimate short-term failure pressure as determined by ASTM D1599.

8.8.2 Modulus of Elasticity

The modulus of elasticity for fiberglass pipe is different in the axial and the hoop directions because the pipe is an anisotropic composite material. Also, the tensile, bending, and compressive moduli may differ significantly, thus it is important to use the correct value. The moduli depend on the type of resin, amount of glass, and orientation of the glass filaments. Precise values for the moduli for specific conditions of loading and temperature should come from the manufacturer. Typical values are often obtained by drawing a tangent to the stress–strain curve at the point equal to one fourth of ultimate failure load. The moduli may also vary with temperature.

8.8.3 Allowable Tensile or Compressive Loads

Typically, the allowable design stress is 25% of the ultimate short-term failure loads. These stress values can be used with the minimum reinforced wall thickness (area) to calculate the allowable maximum loads.

8.8.4 Bending Loads

Ultimate beam stress is determined by using a simple beam with a concentrated load applied to the center to achieve short-term failure. The allowable design stress is then established by application of at least an 8:1 factor of safety to the ultimate failure value. The 8:1 factor is selected to compensate for combined loading that occurs in pressure piping applications.

The bending modulus is determined from a test by measuring mid-span deflections of a simply supported beam with a uniformly distributed load over time, usually not less than 6 weeks. Allowable bending stress and the bending modulus of elasticity may vary with temperature. Values must account for the temperature extremes expected to occur in the piping application under consideration.

8.8.5 Poisson's Ratio

Because fiberglass piping is an anisotropic material, Poisson's ratio varies depending on loading conditions. For example, Poisson's ratio in the transverse (hoop) direction resulting from the axial loading is not the same as Poisson's ratio in the axial direction resulting from transverse (hoop) loading.

8.8.6 Vacuum or External Pressure

Fiberglass pipe can convey materials under vacuum. The ability of fiberglass pipe to resist collapse pressure depends on the pipe stiffness, which is a function of the pipe size, method of manufacture, ratio of diameter to wall thickness, and the raw materials used.

The external pressure resistance of fiberglass pipe may be determined by testing in accordance with ASTM D2924. This standard test method identifies two short-term failure pressures:

- Buckling failure pressure—the external gauge pressure at which buckling occurs.

- Compressive failure pressure—the maximum external gauge pressure that the pipe will resist without transmission of fluid through the wall.

Scaling constants are used to relate the test data to pipe sizes not tested.

Typically, the manufacturer's recommended value for collapse pressure is 33% to 75% of the pipe ultimate short-term external failure pressure (ASTM D2924). The manufacturer's recommended values should be used for design purposes.

8.8.6.1 Buckling scaling constant.

$$K = \frac{P_{ext}}{(E_h)(r/t)^3} \qquad K = \frac{P_{ext} \times 10^{-6}}{(E_h)(r/t)^3} \qquad (8\text{-}15)$$

Where:
K = buckling scaling constant
P_{ext} = external collapse pressure, psig
E_h = circumferential modulus, psi
r = mean radius, in.
t = minimum reinforced wall thickness, in.

Where:
K = buckling scaling constant
P_{ext} = external collapse pressure, kPa
E_h = circumferential modulus, GPa
r = mean radius, mm
t = minimum reinforced wall thickness, mm

8.8.6.2 Compressive scaling constant.

$$C = \frac{P_c(\text{OD} - t)}{2t} \qquad C = \frac{P_c(\text{OD} - t)}{2t} \qquad (8\text{-}16)$$

Where:
C = compressive scaling constant
P_c = compressive failure pressure, psi

Where:
C = compressive scaling constant
P_c = compressive failure pressure, N/m^2

120 FIBERGLASS PIPE DESIGN

8.9 DESIGN EXAMPLES

To assist in understanding the use of the various equations in this chapter, following are several design example calculations.

Design Example Inputs

		in.-lb Units		SI Units	
Parameter	Symbol	Unit	Value	Unit	Value
Pipe outside diameter	OD	in.	2.375	mm	60
Pipe inside diameter	ID	in.	2.235	mm	56
Pipe reinforced wall thickness	T	in.	0.07	mm	2
Coefficient of thermal expansion	C_t	in./in./°F	1.09×10^{-5}	mm/mm/°C	1.96×10^{-5}
Axial compressive modulus at minimum temperature	E_c	psi	1.3×10^6	GPa	8.9
Axial compressive modulus at maximum temperature	E_c	psi	0.6×10^6	GPa	4.1
Axial bending modulus at minimum temperature	E_b	psi	2.2×10^6	GPa	15.2
Axial bending modulus at maximum temperature	E_b	psi	1.3×10^6	GPa	8.9
Axial tensile modulus	E_t	psi	1.72×10^6	GPa	11.9
Allowable bending stress	σ_b	psi	1,850	MPa	12.8
Installation temperature	—	°F	75	°C	24
Maximum operating temperature	—	°F	200	°C	93
Minimum operating temperature	—	°F	35	°C	2
Temperature change	T_c	°F	calc	°C	calc
Maximum temperature rating	T_r	°F	210	°C	99
Material variation factor	S_f	—	0.9	—	0.9

NOTE: The values in the above table are not direct conversions from one unit system to the other and in many cases rounded values are used for ease of presentation.

in.-lb Unit Examples	SI Unit Examples
Example 1: Determine the length change per unit length for a temperature change of 60°F.	Example 1: Determine the length change per unit length for a temperature change of 33°C.
Use Eq 8-1.	Use Eq 8-1.
$L_c = 12(C_t)(L)(T_c)$ $= 12(1.09 \times 10^{-5})(1)(60)$ $= 0.0078$ in./ft	$L_c = 1{,}000(C_t)(L)(T_c)$ $= 1{,}000(1.96 \times 10^{-5})(1)(33)$ $= 0.65$ mm/m
Example 2: Determine the thermal end loads at maximum and minimum operating temperatures.	Example 2: Determine the thermal end loads at maximum and minimum operating temperatures.
Use Eq 8-2.	Use Eq 8-2.
$E_L = (C_t)(E)(A)(T_c)$	$E_L = (C_t)(E)(A)(T_c) \times 10^9$

ABOVEGROUND PIPE DESIGN AND INSTALLATION   121

Expansion temperature change,
$T_c = 200 - 75 = 125°F$

Contraction temperature change,
$T_c = 75 - 35 = 40°F$

Cross-sectional area,
$A = \pi/4(OD^2 - ID^2)$
$A = 3.14/4(2.375^2 - 2.235^2)$
$A = 0.507$ in.^{2}

Expansion end load =
$(1.09 \times 10^{-5})(1.3 \times 10^6)(0.507)(125)$
use compressive mod for expansion
$= 898$ lb

Contraction end load =
$(1.09 \times 10^{-5})(1.72 \times 10^6)(0.507)(40)$
use tensile mod for expansion
$= 380$ lb

Example 3: Calculate maximum allowable guide spacing.

Use Eq 8-3.

$$L_G = \sqrt{\frac{(\pi^2)(E_b)(I)}{144(C_t)(A)(T_c)(E_c)}}$$

$I = \pi/64 \,(OD^4 - ID^4)$
$= 3.14/64(2.375^4 - 2.235^4)$
$= 0.337$ in.^{4}

Calculate E_b/E_c at the minimum and maximum temperatures.

Min. $2.2 \times 10^6/1.3 \times 10^6 = 1.69$
Max. $1.3 \times 10^6/0.6 \times 10^6 = 2.17$

Use the lesser ratio as conservative and use the maximum temperature change of 125°F.

$$L_G = \frac{(3.14)^2(1.69)(0.337)}{(144)(1.09 \times 10^{-5})(0.507)(125)}$$

$= 7.5$ ft

Expansion temperature change,
$T_c = 93 - 24 = 69°C$

Contraction temperature change,
$T_c = 24 - 2 = 22°C$

Cross-sectional area,
$A = \pi/4(OD^2 - ID^2)$
$A = 3.14/4(60^2 - 56^2) \times 10^{-6}$
$A = .00036$ m^2

Expansion end load =
$(1.96 \times 10^{-5})(8.9)(0.00036)(69) \times 10^9$
use compressive mod for expansion
$= 4,333$ N

Contraction end load =
$(1.96 \times 10^{-5})(11.9)(0.00036)(22) \times 10^9$
use tensile mod for expansion
$= 1,847$ N

Example 3: Calculate maximum allowable guide spacing.

Use Eq 8-3.

$$L_G = \sqrt{\frac{(\pi^2)(E_b)(I)}{(C_t)(A)(T_c)(E_c)}}$$

$I = \pi/64 \,(OD^4 - ID^4) \times 10^{-12}$
$= 3.14/64(60^4 - 56^4)$
$= 0.153 \times 10^{-6}$ m^4

Calculate E_b/E_c at the minimum and maximum temperatures.

Min. $15.2/8.9 = 1.71$
Min. $8.9 / 4.1 = 2.17$

Use the lesser ratio as conservative and use the maximum temperature change of 69°C.

$$L_G = \frac{(3.14)^2(1.71)(0.153 \times 10^{-6})}{(1.96 \times 10^{-5})(0.00036)(69)}$$

$= 2.3$ m

Example 4: Calculate critical buckling force.

Use Eq 8-4.

$$P_{cr} = \frac{\pi^2(E_c)(I)(S_f)}{L_G^2}$$

$$= \frac{3.14^2(1.3 \times 10^6)(0.337)(0.9)}{(7.5 \times 12)^2}$$

$$= 481 \text{ lb}$$

Example 5: Calculate the length of expansion loop "A" leg assuming maximum length change of 4.0 in.

Use Eq 8-5.

$$L_A = \sqrt{\frac{(K)(L_c)(E_b)(\text{OD})}{(144)(\sigma_b)}}$$

Assume a nonguided cantilever beam and use $K = 0.75$

$$= \frac{(0.75)(4))(1.3 \times 10^6)(2.375)}{144(1,850)}$$

$$= 5.9 \text{ ft}$$

If, as typically assumed, leg "B" is taken as half of leg "A," then "B" = 5.9/2 = 2.95 ft.

Example 6: Calculate the length from directional change to the first secure hanger.

Use Eq 8-7.

$$L_{sh} = \sqrt{\frac{1.5(L_c)(E_b)(\text{OD})}{(144)(\sigma_b)}}$$

$$= \frac{1.5(4.0)(1.3 \times 10^6)(2.375)}{144(1,850)}$$

$$= 8.3 \text{ ft}$$

Example 4: Calculate critical buckling force.

Use Eq 8-4.

$$P_{cr} = \frac{\pi^2(E_c)(I)(S_f) \times 10^9}{L_G^2}$$

$$= \frac{3.14^2(8.9)(0.153 \times 10^{-6})(0.9) \times 10^9}{2.3^2}$$

$$= 2,284 \text{ N}$$

Example 5: Calculate the length of expansion loop "A" leg assuming maximum length change of 100 mm.

Use Eq 8-5.

$$L_A = \sqrt{\frac{(K)(L_c)(E_b)(\text{OD})}{(1,000)(\sigma_b)}}$$

Assume a nonguided cantilever beam and use $K = 0.75$

$$= \frac{(0.75)(100)(8.9)(60)}{1,000(12.8)}$$

$$= 1.77 \text{ m}$$

If, as typically assumed, leg "B" is taken as half of leg "A," then "B" = 1.77/2 = 0.89 m.

Example 6: Calculate the length from directional change to the first secure hanger.

Use Eq 8-7.

$$L_{sh} = \sqrt{\frac{1.5(L_c)(E_b)(\text{OD})}{(1,000)(\sigma_b)}}$$

$$= \frac{1.5(100)(8.9)(60)}{1,000(12.8)}$$

$$= 2.5 \text{ m}$$

Example 7: Calculate the allowable unsupported span length.

Use Eq 8-8.

$$L_s = \sqrt[4]{\frac{(d_m)(E_b)(I)}{0.013(W)}}$$

Assume an allowable midspan deflection (d_m) of 0.5 in., as is typical for fiberglass pipe, and a pipe weight (W_p) of 0.4 lb/ft for a 1-ft length.

$$V_p = \frac{\pi(\text{ID})^2(L)}{4(12)^2}$$

$$= \frac{3.14(2.235)^2(1)}{4(12)^2}$$

$$= 0.0272 \text{ ft}^3/\text{ft}$$

$W_f = (V_p)p$
$\quad = 0.0272 \,(62.4)$
$\quad = 1.69 \text{ lb/ft}$

$W = W_f + W_p$
$\quad = 1.69 + 0.4$
$\quad = 2.09 \text{ lb/ft}$

$$L_s = \sqrt[4]{\frac{0.5(1.3 \times 10^6)(0.337)}{0.013(2.09/12)}}$$

$$= 99.2 \text{ in.} = 8.26 \text{ ft}$$

Example 8: Calculate the minimum allowable bending radius.

Use Eq 8-12.

$$R_m = \frac{(E_b)(\text{OD})}{24\sigma_b}$$

$$= \frac{(2.2 \times 10^6)(2.375)}{24(1,850)}$$

$$= 117.7 \text{ ft}$$

Example 7: Calculate the allowable unsupported span length.

Use Eq 8-8.

$$L_s = \sqrt[4]{\frac{(d_m)(E_b)(I) \times 10^9}{13(W)}}$$

Assume an allowable midspan deflection (d_m) of 12.5 mm, as is typical for fiberglass pipe, and a pipe weight (W_p) of 0.6 kg/m (5.9 N/m) for a 1-m length.

$$V_p = \frac{\pi(\text{ID})^2(L)}{4(1 \times 10^6)}$$

$$= \frac{3.14(56)^2(1)}{4(1 \times 10^6)}$$

$$= 0.0025 \text{ m}^3/\text{m}$$

$W_f = (V_p)p$
$\quad = 0.0025 \,(9,800)$
$\quad = 24.5 \text{ (N/m)}$

$W = W_f + W_p$
$\quad = 24.5 + 5.9$
$\quad = 30.4 \text{ (N/m)}$

$$L_s = \sqrt[4]{\frac{12.5(8.9)(0.153 \times 10^6) \times 10^9}{13(30.4)}}$$

$$= 2.56 \text{ m}$$

Example 8: Calculate the minimum allowable bending radius.

Use Eq 8-12.

$$R_m = \frac{(E_b)(\text{OD})}{2\sigma_b}$$

$$= \frac{(15.2)(60)}{2(12.8)}$$

$$= 35.6 \text{ m}$$

Example 9: Determine the maximum heat tracing temperature allowed to maintain a 95°F temperature inside a pipe with a maximum temperature rating of 210°F and conveying a fluid with a maximum chemical resistance temperature of 100°F.

Criteria 1—Average wall temperature not to exceed maximum temperature rating.

Use Eq 8-13.

A_t = average wall temperature
$= (T_i + T_t)/2 = 210$
$= (95 + T_t)/2 = 210$
$T_t = 325°F$

Criteria 2—Maximum heat tracing temperature not to be more than 100°F above the maximum rated temperature of the pipe.

Use Eq 8-14.

$T_t = T_R + 100°F$
$= 210 + 100 = 310°F$

Use the lesser of the two determinations or in this case a maximum heat trace temperature of 310°F.

Criteria 3—The maximum recommended chemical resistance temperature of the pipe must not be exceeded at the inside pipe wall.

For this example in flowing conditions, the inside pipe wall will not exceed the recommended maximum chemical resistance temperature of 100°F. However, if stagnant conditions could be anticipated, the heat tracing temperature should be limited to 100°F.

Example 9: Determine the maximum heat tracing temperature allowed to maintain a 35°C temperature inside a pipe with a maximum temperature rating of 99°C and conveying a fluid with a maximum chemical resistance temperature of 38°C.

Criteria 1—Average wall temperature not to exceed maximum temperature rating.

Use Eq 8-13.

A_t = average wall temperature
$= (T_i + T_t)/2 = 99$
$= (35 + T_t)/2 = 99$
$T_t = 163°C$

Criteria 2—Maximum heat tracing temperature not to be more than 100°F above the maximum rated temperature of the pipe. (NOTE: The 100°F is a Δ temperature, which must be converted to metric using 0.56°C/°F.)

Use Eq 8-14.

$T_t = T_R + 56°C$
$= 99 + 56 = 155°C$

Use the lesser of the two determinations or in this case a maximum heat trace temperature of 155°C.

Criteria 3—The maximum recommended chemical resistance temperature of the pipe must not be exceeded at the inside pipe wall.

For this example in flowing conditions, the inside pipe wall will not exceed the recommended maximum chemical resistance temperature of 38°C. However, if stagnant conditions could be anticipated, the heat tracing temperature should be limited to 38°C.

AWWA MANUAL M45

Chapter 9

Joining Systems, Fittings, and Specials

9.1 INTRODUCTION

Several types of joining systems are available for use with fiberglass pressure pipe. Many of the systems permit joint angular deflection. Some joining systems may be designed to resist longitudinal thrust forces. Fittings and specials are available in a range of styles and configurations and are fabricated using a number of different manufacturing methods.

9.2 FIBERGLASS PIPE JOINING SYSTEMS CLASSIFICATION

There are two general joint classifications: unrestrained and restrained.

9.2.1 Unrestrained Pipe Joints

Unrestrained pipe joints can withstand internal pressure but do not resist longitudinal forces. They rely on elastomeric gaskets to provide the seal. Typically, these joints can be disassembled without damage.

9.2.1.1 Fiberglass couplings or bell-and-spigot joints. These joints use an elastomeric seal located in a groove on the spigot or in the bell as the sole means to provide fluid tightness.

9.2.1.2 Mechanical coupling joint. These joints use mechanically energized elastomeric gasket seals to join two pieces of pipe. The mechanical coupling technique applies to plain end pipe.

9.2.2 Restrained Pipe Joints

Restrained pipe joints can withstand internal pressure and resist longitudinal forces.
Joints that may later be disassembled without damage include:

- coupling or bell-and-spigot with a restraining device
- flange
- mechanical

Joints that cannot be disassembled without damage or cutting apart include:

- butt and wrap
- wrapped bell-and-spigot
- bonded bell-and-spigot

9.3 GASKET REQUIREMENTS

Gaskets used with fiberglass pipe joining systems should conform to the requirements of ASTM F477. The gasket material composition must be selected to be compatible with the intended environment.

9.4 JOINING SYSTEMS DESCRIPTION

In this section, many of the joining systems available with fiberglass pressure pipe are described; however, the details of every type of joining system available are not included. Versatility of manufacture permits differences in configuration and geometry while meeting performance requirements. Users should contact the pipe manufacturer to obtain specific details on joints and joint performance.

9.4.1 Adhesive-Bonded Joints

Three types of adhesive-bonded joints are available:

- a joint using a tapered bell and a tapered spigot (Figure 9-1),
- a straight bell and straight spigot joint (Figure 9-2), and
- a joint using a tapered bell and a straight spigot (Figure 9-3).

Adhesive-bonded joints are generally available for pipe up through 16-in. (400-mm) diameter.

9.4.2 Reinforced-Overlay Joints

The butt-and-wrap joint typically consists of two squared pipe ends that have been prepared for joining by roughening the outside surface in the joint area. The pipes are then abutted end to end, aligned on the same centerline, and the joint overwrapped with layers of resin-impregnated glass fiber materials. Each layer becomes increasingly wider to provide a buildup that accommodates internal pressure and longitudinal forces. Basic joint construction is shown in Figure 9-4, with the finished joint illustrated in Figure 9-5. A variation of this joint is illustrated in Figure 9-6, in which the pipe ends are tapered. Bell-and-spigot joints are sometimes overlaid as shown in Figure 9-7. In this system the bell aids in alignment during the overlay operation. Internal overlays are also used to improve joint performance but are generally only possible on larger diameter pipe that allows the installer to work inside the pipe during installation.

9.4.3 Gasket-Sealed Joints

9.4.3.1 Bell-and-spigot. Figures 9-8 and 9-9 illustrate a bell-and-spigot gasketed joint using a single-gasket design. Figures 9-10 and 9-11 illustrate a bell-and-spigot gasketed joint using double-gasket design. The double-gasket design is generally only

JOINING SYSTEMS, FITTINGS, AND SPECIALS   127

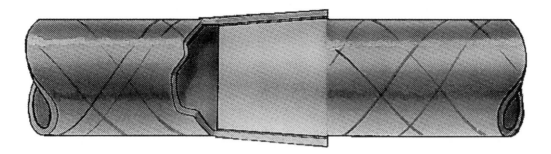

Source: Smith Fiberglass Products Inc., Little Rock, Ark.

Figure 9-1   Tapered bell-and-spigot joint

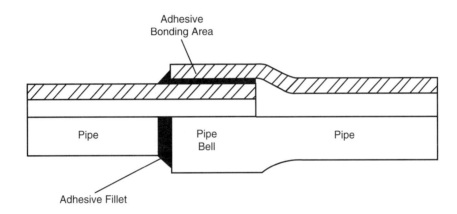

Figure 9-2   Straight bell and straight spigot joint

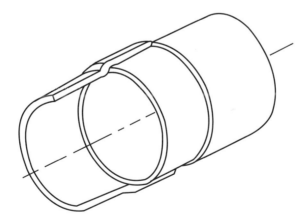

Figure 9-3   Tapered bell and straight spigot joint

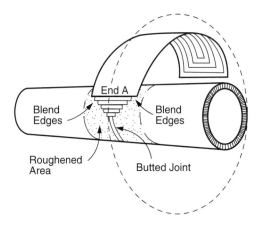

Reprinted with permission from Fiberglass Pipe Handbook, *Fiberglass Pipe Institute, New York, N.Y.*

Figure 9-4   Overlay joint construction

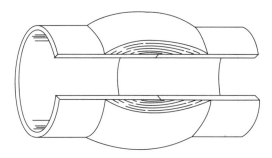

Source: Smith Fiberglass Products Inc., Little Rock, Ark.

Figure 9-5   Overlay joint

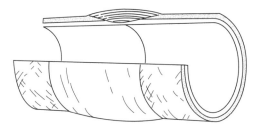

Reprinted with permission from Fiberglass Pipe Handbook, *Fiberglass Pipe Institute, New York, N.Y.*

Figure 9-6   Tapered ends overlay joint

used with larger diameter pipe. By inserting ports in the spigot between the two gaskets, a test of the sealing integrity of the gaskets can be conducted immediately after assembly using hydrostatic or pneumatic pressure.

9.4.3.2 Coupling. Figures 9-12 and 9-13 show two styles of gasketed coupling joints. The joint in Figure 9-12 uses a gasket mechanically bonded or molded in the coupling. Figure 9-13 shows a coupling with gaskets retained in grooves.

9.4.3.3 Restrained gasketed joints. None of the gasketed joints shown in Figures 9-8 through 9-13 provide longitudinal restraint, although they can be modified

JOINING SYSTEMS, FITTINGS, AND SPECIALS   129

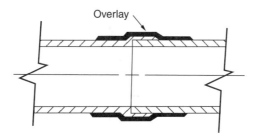

Reprinted with permission from Fiberglass Pipe Handbook, *Fiberglass Pipe Institute, New York, N.Y.*

Figure 9-7   Bell-and-spigot overlay joint

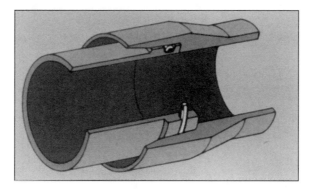

Source: Hobas Pipe USA Inc., Houston, Texas.

Figure 9-8   Single-gasket bell-and-spigot joint

Source: Price Brothers Company, Dayton, Ohio.

Figure 9-9   Single-gasket spigot

in a variety of ways to do so. Figure 9-14 illustrates a bell-and-spigot joint with a gasket and restraining elements. The restraining element is a mechanically loaded locking ring designed to expand and allow the spigot to enter the bell and then contract to lock on a shoulder of the spigot outside diameter. Figure 9-15 illustrates a coupling joint with a pair of gaskets and restraining elements. The shape and the material used for the restraining element can vary. Both metallic and shear-resistant plastic

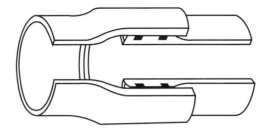

Source: Smith Fiberglass Products, Little Rock, Ark.

Figure 9-10 Double-gasket bell-and-spigot joint

Source: Price Brothers Company, Dayton, Ohio.

Figure 9-11 Double-gasket spigot

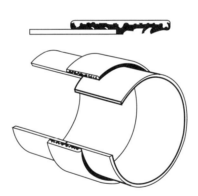

Source: Hobas Pipe USA Inc., Houston, Texas.

Figure 9-12 Gasketed coupling joint

materials are used for this device. Figure 9-16 illustrates a bell-and-spigot joint with a gasket and a threaded connection joint restraining element. An advantage of many of the restrained joints is that they can be disassembled for removal or repair.

9.4.4 Mechanical Joints

There are numerous mechanical joints available for use with fiberglass pipe, including flanges, threaded joints, and commercially available proprietary joints. Pressure-rated flanges are common in the installation of all sizes of fiberglass pressure pipe. Fiberglass

JOINING SYSTEMS, FITTINGS, AND SPECIALS   131

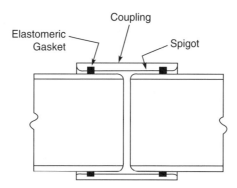

Source: Flowtite Technology, Sandefjord, Norway.

Figure 9-13   Gasketed coupling joint—cross section

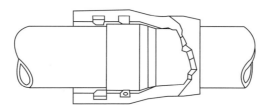

Source: Smith Fiberglass Products Inc., Little Rock, Ark.

Figure 9-14   Restrained-gasketed bell-and-spigot joint

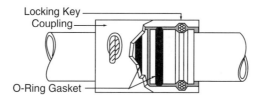

Source: Smith Fiberglass Products Inc., Little Rock, Ark.

Figure 9-15   Restrained-gasketed coupling joint

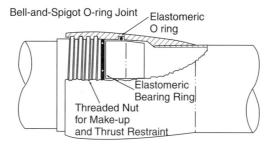

Figure 9-16   Restrained-gasketed threaded bell-and-spigot O-ring joint

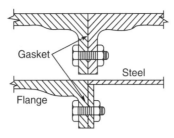

Source: Old Hope Corguard Inc. (former subsidiary of Price Brothers Company, Dayton, Ohio).

Figure 9-17 Fiberglass flange to fiberglass and steel flange joint

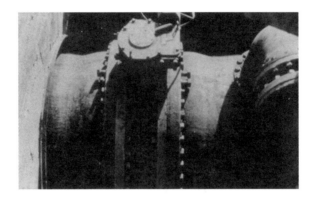

Source: Price Brothers Company, Dayton, Ohio.

Figure 9-18 Fiberglass flanges to flanged steel valve connection

flanges have bolting dimensions consistent with standard ANSI/ASME pressure classes of bolted flanges. Fiberglass flanges are produced by hand lay-up, filament winding, and compression molding.

Project conditions often dictate mating a fiberglass flange with a metallic flange on a pump, valve, or metallic pipe. Figure 9-17 depicts a fiberglass flange to fiberglass flange joint and a fiberglass flange to steel flange joint. Figure 9-18 shows the joining of fiberglass flanges to steel flanges to complete a valve connection.

Gaskets used with fiberglass flanges may be flat-faced or O-rings contained in a groove in the flange face (see Figure 9-19). The use of O-ring seals has been found to be very effective, particularly for large diameters, because positive seal is obtained without excessive bolt torque.

Figure 9-20 shows one common mechanically coupled joint where the seal is accomplished on the outside surface of the pipe. This type of joint does not accommodate longitudinal forces. Care must be taken to not overtorque this type of mechanical joint because excessive torque can damage some fiberglass pipe.

9.5 ASSEMBLY OF BONDED, THREADED, AND FLANGED JOINTS

Bonded, threaded, and flanged fiberglass pipe joints require the use of techniques and equipment that may be considerably different than those used with other piping materials. Although the pipe manufacturer's instructions must always be followed, a brief general overview is given in the following sections.

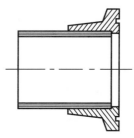

Reprinted with permission from Fiberglass Pipe Handbook, *Fiberglass Pipe Institute, New York, N.Y.*

Figure 9-19   Fiberglass flange with grooved face for O-ring seal

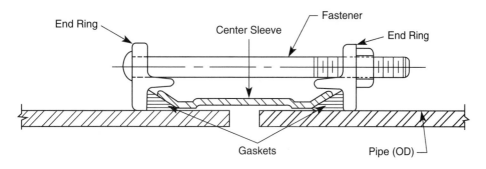

Figure 9-20   Mechanical coupling joint

9.5.1   Layout and Preparation

Installation crew members must be familiar with the installation procedures provided by the manufacturer.

Inspection of the pipe and fittings for damage that may have occurred during handling is important. Proper storage and handling procedures are discussed in chapter 10 and provided by the manufacturers.

The crew size requirement varies from one type of installation to another. A typical crew for 2 to 4 in. (50 to 100 mm) diameter pipe installations is two or three members, while installations involving large diameters can require crews of four or more members.

9.5.2   Tool and Equipment Requirements

Tool and equipment requirements vary with pipe size and type of joint; however, the following are general guidelines.

For cold weather installations, heating devices such as electric heating collars, heated portable buildings (plastic huts), hot air blowers, etc., are necessary to ensure proper installation of bonded joints.

Machining equipment such as tapering tools, disk sanders, etc., are required for end preparation on bonded joint systems. Specialized machines, such as tapering tools, are often available from the manufacturer. Disk grinders, belt sanders, and other more common equipment are generally supplied by the installer.

Pipe cutting equipment usually consists of fine-tooth saws and/or saws with carbide grit abrasive blades. Saw blades and hole saws typically used for wood are not suitable; however, blades used for masonry and/or tiles are usually abrasive-type blades that will be suitable for fiberglass pipe. NOTE: Cutting and/or grinding operations can generate dust or cutting chips that are irritating to the skin, upper respiratory tract, and eyes. Because these materials are irritating, good ventilation for the installation crew should be used to prevent overexposure. A nuisance dust breathing filter should be used when working in areas where wind and dust are present. Tool operators should wear heavy cotton clothing, including long-sleeve shirts, which protect the skin from the dust. Eye protection is often required for tool operators. Contact your local regulatory agency or Occupational Safety and Health Administration (OSHA) office for specific requirements on the use of respirators, protective clothing, and any additional safeguards.

Pipe chain vises and pipe stands are designed for metal piping. Therefore, it is necessary to provide protective pads, such as rubber cushions, to protect the pipe from point loading and/or impact damage. Protective pads are sometimes required when using come-a-longs or other tools that can create bearing and/or point loading damage.

For threaded joints, special wrenches and/or strap wrenches are recommended by most manufacturers. CAUTION: Improper use of strap wrenches can cause point loading.

Some tools can be used with a power drive, such as a Rigid 700 or a Rigid 300. The contractor may have to obtain a different adapter for the power drives. For example, threaded adapters used by many contractors are not used for fiberglass pipe. A typical adapter consists of a 1-in. (25-mm) drive socket that fits a $^{15}/_{16}$-in. (24-mm) square drive.

Miscellaneous equipment such as a wrap-around, felt tip marking pens, hammers (metal and rubber), and adjustable pipe stands are also required for installation.

9.5.3 Bonded Joint Assembly

Because there are many different types of joints available, detailed assembly instructions are beyond the scope of this manual. It is essential that manufacturer's instructions be obtained for each type of joint being installed. Following are general guidelines.

Clean bonding surfaces are required for proper adhesion of adhesives and/or resins. In some cases, a cleaning operation, such as washing and using cleaning solvent, is recommended. In all cases, avoid contamination that will leave dirt, oil, grease, fingerprints, etc., on surfaces that require adhesive or resin applications. Thoroughly mix the adhesive or resin and follow safety precautions that are included with the materials. In most cases, the adhesive materials are preweighed and it is not possible to "split a kit."

Shelf (storage) life and working (pot) life will vary from one type of resin to another. If the mixture is setting up too fast or not at all, consult the manufacturer to determine the best storage conditions, shelf life, and typical working life.

End preparation varies for the different joints. However, a clean, machined surface is generally required for application of adhesive or resin. The machining operation may involve sanding or grinding with special tools. For general sanding operations, a coarse grit (24 or less) sandpaper is better than a fine grit.

Application of adhesives and/or resins normally requires a "wetting" process (i.e., the materials should be applied in a manner that increases the penetration—and bonding—of the resins to the substrate), for example, using pressure on a paintbrush to apply resin to a machined surface.

Cure times vary, and not all mixtures are properly cured when they have set up (or are hard to the touch). The proper mixing and curing procedures from the manufacturer must be followed to ensure maximum physical strength and proper chemical resistance for the system. CAUTION: If a mixture becomes warm and starts to cure in the container, discard it immediately. Do not use this material to assemble a joint.

In some cases, it is necessary to apply heat to speed up or ensure completion of the curing process. CAUTION: Allow a heated joint to cool until it is comfortable to the touch before any stress is applied to the joint. Any stresses on the pipe due to bending or sagging should be relieved prior to heat cure.

9.5.4 Threaded Joints

Connecting to other systems is typically accomplished with mechanical connections, threaded adapters (National Pipe Threads), reducer bushings (National Pipe Threads), grooved adapters, or flanges. Flange patterns are usually 150 lb (68 kg) or 300 lb (136 kg) bolt circle for small-diameter systems and 125 lb (57 kg) bolt circle for larger diameter systems (above 24 in. [600 mm]).

Before making up threaded connections, inspect the threads. Do not use fittings with damaged threads. Inspect all metal threads. Remove any burrs and reject metal threads that have notches (grooves) that are near the end of the threads. The quality of metal threads is a concern when mating to fiberglass threads that require a low torque level. The quality of the metal threads will often have little or no effect on metal-to-metal connections because the use of additional torquing force may seal a leak. Fiberglass-to-steel connections are more likely to leak if the steel threads are in poor condition.

Unless a union is used, threaded adapters should be threaded into the other system before assembly of the fiberglass piping. Best results will be obtained using a strap wrench and a solvent-free, soft-set, nonmetallic thread lubricant. If thread sealing tapes are used, avoid improper installation of the tape, such as using thick layers of tape, to prevent damage to the fiberglass threads. In all cases, tighten the fiberglass threads as if they were brass or other soft material.

9.5.5 Flanged Joints

Most fiberglass flanges are designed for use against a flat surface; therefore, it may be necessary to use spacers or reinforcement (back-up) rings for connections to metal flanges, valves, pumps, etc. Fiberglass flanges require the use of flat washers on all bolts and nuts. In most cases, the type of gasket is specified by the manufacturer and may have a flat, "O"-ring, or other gasket configuration.

9.5.6 Safety Precautions

Testing with air or gas is not recommended because of the safety hazards involved. The light weight, flexibility, and elasticity of fiberglass pipe create conditions that are different from those present with steel pipe. If a catastrophic failure occurs in a fiberglass system, the system would be subject to considerable whipping and other shock-induced conditions due to the sudden release of stored energy. The recommended procedure is to conduct a hydrostatic pressure test.

9.6 FITTINGS AND SPECIALS

Fiberglass fittings and specials are available over a wide range of diameters, pressures, and configurations. Fittings and specials are made by compression molding, filament winding, cutting and mitering, and contact molding.

9.6.1 Compression Molding

Compression molding is generally used for fittings up to 16 in. (400 mm) diameter. Figures 9-21 and 9-22 illustrate the range of configurations available for use with

Figure 9-21 Compression molded fittings

Figure 9-22 Flanged compression molded fittings

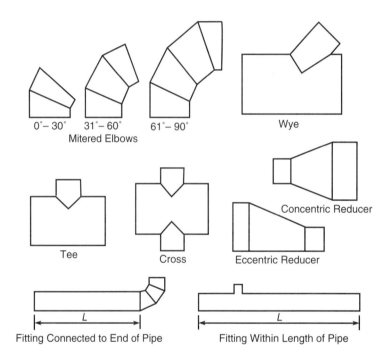

Figure 9-23 Mitered fitting configurations

plain-end or flanged joints, for pressure applications of less than 500 psi (3,447 kPa). In this process, a weighed glass/resin mixture is placed in a multipiece mold. The mold pieces are then held together with high pressure while the temperature is increased to cause curing. Molded fittings are cost-effective for low-pressure, small-diameter applications where a large number of fittings are required.

9.6.2 Filament Winding

Filament winding can produce fittings with higher mechanical strength than is possible with molded fittings. In this process, resin-impregnated glass fibers are wound

Source: Price Brothers Company, Dayton, Ohio.

Figure 9-24 Mitered fitting

Source: Price Brothers Company, Dayton, Ohio.

Figure 9-25 Mitered fitting fabrication

onto a fitting jig. The process may also include the use of woven rovings and/or glass mat. After winding and curing, the fitting is removed from the jig for postproduction processing.

9.6.3 Cut and Miter Process

The cut and miter process is extremely versatile for making the full range of diameters, standard and special shapes, and custom-designed fittings. Figure 9-23 shows a sampling of the fittings that are routinely made from cut and mitered sections. Fabrication of the fittings and specials starts with the production of pipe that is cut and assembled into the desired configuration.

Cut and mitered fittings also can be made by cutting pipe sections to the desired form. Pieces are joined together with contact molding techniques using chopped strand and woven roving reinforcement. Surface preparation before bonding is essential to ensure good adhesion between surfaces and the contact molded laminate. The fitting should resist the same loading conditions as the pipe.

Figures 9-24 through 9-28 show a variety of fiberglass fittings and specials during fabrication and on installation sites.

9.6.4 Contact Molding

Contact molding (including spray-up) may be used to produce fittings directly.

9.7 SERVICE LINE CONNECTIONS

Service line connections are typically made using tapping saddles. Consult individual pipe manufacturers for procedures applicable to specific products.

REFERENCE

American Society for Testing and Materials. ASTM F477, *Standard Specification for Elastomeric Seals (Gaskets) for Joining Plastic Pipe*. West Conshohocken, Pa.: American Society for Testing and Materials.

138 FIBERGLASS PIPE DESIGN

Source: Price Brothers Company, Dayton, Ohio.

Figure 9-26 Mitered fittings

Source: Price Brothers Company, Dayton, Ohio.

Figure 9-27 Mitered fitting field fabrication

Source: Price Brothers Company, Dayton, Ohio.

Figure 9-28 Fittings field assembly

AWWA MANUAL M45

Chapter 10

Shipping, Handling, Storage, and Repair

10.1 INTRODUCTION

Fiberglass pipes encompass a wide range of diameters (1 in. to 144 in. [25 mm to 3,600 mm]) with an equally wide range of wall thicknesses (from less than 0.1 in. to more than 3 in. [3 mm to 80 mm]). Furthermore, the wall laminate constructions and characteristics vary sufficiently to exhibit significantly different behaviors. Due to this wide variation in design and material characteristics, the requirements for acceptable shipping, handling, and storage are also somewhat variable. Consult the manufacturer for procedures specific to its products.

Despite the many differences, there are also numerous similarities and therefore several procedures that are typical and prudent for all fiberglass pipes. These procedures and suggestions should be used in conjunction with the pipe manufacturer's instructions. The handling requirements for fiberglass pipe are similar to those for all types of pipe.

10.2 SHIPPING

Preparation for shipping should protect the pipe wall and joining ends from damage and should be acceptable to the carrier, the manufacturer, and the purchaser.

Ship pipe on flatbed trucks supported on flat timbers or cradles (see Figure 10-1). A minimum of two supports located at the pipe quarter points is typical. Timber supports should contact only the pipe wall (no joint surfaces). No bells, couplings, or any other joint surface should be permitted to contact the trailer, supports, or other pipe. The timber supports must be of sufficient width to avoid point loading. Chock the pipes to maintain stability and separation. To ensure that vibrations during transport do not cause abrasion damage, do not allow pipes to contact other pipes. Strap the pipe to the vehicle over the support points using pliable straps or rope without

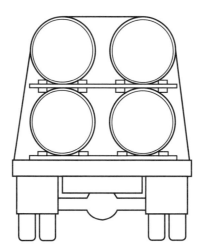

Source: Flowtite Technology, Sandefjord, Norway.

Figure 10-1 Pipe shipment by truck

deforming the pipe. Bulges, flat areas, or other abrupt changes in pipe curvature are not permitted. Stack heights to the legal limits are typically acceptable.

Inspect the pipe upon receipt at the job site for loss or damage sustained in transit. Exterior inspection is usually sufficient; however, impact to the pipe exterior can cause interior cracking with little or no visible damage to the pipe exterior. Therefore, interior inspection at the location of exterior scrapes may be helpful when pipe size permits such an inspection. If the load has shifted or exhibits broken packaging, inspect each piece both internally and externally. Reinspect the pipe just prior to installation. If any imperfections or damage are found, contact the supplier for recommendations concerning repair and replacement. NOTE: Do not use pipe that appears damaged or defective. If in doubt, do not use. If it is necessary to transport pipes at the job site, it is best to use the original shipping dunnage.

10.3 HANDLING

Manufacturers' instructions regarding use of slings, spreader bars, or other handling devices should be followed. Lift pipe sections with wide fabric straps, belts, or other pliable materials. Do not allow the straps to deform the pipe. Avoid the use of steel cables, chains, or other materials that may damage the pipe surface. If cables, chains, or forklifts are used, sufficient care, padding, or protection must be used to prevent gouging, cutting, or otherwise damaging the pipe.

Individual pipe sections can usually be lifted with a single sling (see Figure 10-2) if properly balanced, but two slings, as shown in Figure 10-3 (located at the pipe quarter points), make the pipe easier to control. Do not lift pipe with hooks or rope inserted through the pipe ends.

Because fiberglass pipe may be damaged by impact, do not drop or impact the pipe, especially the pipe ends. Pipe should never be thrown or dropped to the ground or set on sharp objects. Repair any damage prior to installation.

Bundles. Smaller pipe (24 in. [600 mm] diameter and less) are often unitized or bundled by the manufacturer, as shown in Figures 10-4 and 10-5. Bundles and unitized loads typically must be handled with a pair of slings (never a single sling). Do not lift a nonunitized stack of pipe as a single unit. Nonunitized stacked pipe must be unstacked and handled individually.

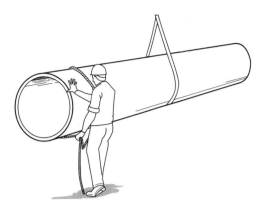

Source: Flowtite Technology, Sandefjord, Norway.

Figure 10-2 Single sling handling

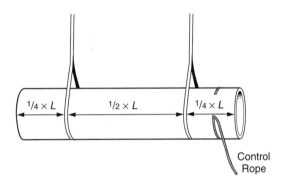

Source: Flowtite Technology, Sandefjord, Norway.

Figure 10-3 Double sling handling

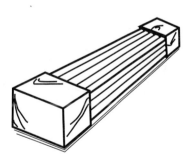

Source: Flowtite Technology, Sandefjord, Norway.

Figure 10-4 Unitized small-diameter bundle

Nested pipe. Nesting smaller pipes inside larger pipes is acceptable. Ensure that the pipes are protected and secured properly to prevent relative motion or damage during shipment. The pipe manufacturer will provide written instructions for shipping, handling, and denesting of pipe. Never lift nested pipe with a single strap; always use two or more straps, as shown in Figure 10-6. Ensure that the lifting straps have the capacity to hold the bundle weight. Denesting is typically accomplished with

142   FIBERGLASS PIPE DESIGN

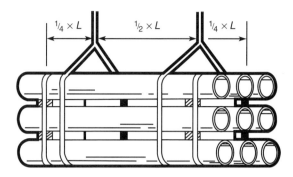

Source: Flowtite Technology, Sandefjord, Norway.

Figure 10-5   Unitized load handling

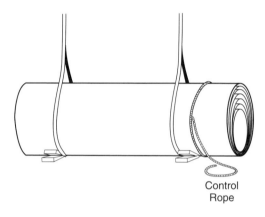

Source: Flowtite Technology, Sandefjord, Norway.

Figure 10-6   Handling nested pipes

three or four fixed cradles that match the outside diameter of the largest pipe in the bundles. Denest beginning with the inside pipe (smallest diameter). The standard denesting procedure is to insert a padded forklift boom, lift slightly to suspend the pipe, and carefully remove it without touching the other pipe (see Figure 10-7). When weight, length, and equipment limitations preclude this method, check with the manufacturer for specific recommendations for removing pipe from the bundle.

10.4   STORAGE

Pipe is generally stored on flat timbers to facilitate placement and removal of lifting slings (see Figure 10-8). The support timbers should be of sufficient width to prevent point loads. Supports that are 4-in. (10-cm) wide are recommended for large-diameter pipe. Pipe should be chocked to prevent rolling in high winds. When stacking, timber supports at the pipe quarter points are best. If available, use the original shipping dunnage for storage. The maximum stack height is typically 8 ft (2.4 m). Consult the manufacturer for maximum storage deflection. Bulges, flat areas, or other abrupt changes in pipe curvature are not permitted. Nylon or hemp rope tie-downs are best. Chain tie-downs must be well padded to prevent damage to the pipe wall.

SHIPPING, HANDLING, STORAGE, AND REPAIR   143

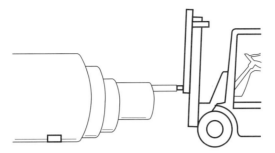

Source: Flowtite Technology, Sandefjord, Norway.

Figure 10-7 Denesting pipes

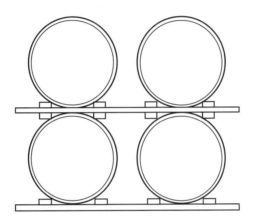

Source: Flowtite Technology, Sandefjord, Norway.

Figure 10-8 Pipe stacking

Rubber ring gaskets should be stored in the shade in the original packaging. They typically must be protected from sunlight, solvents, and petroleum-based greases and oils.

When stored directly on the ground, the pipe weight should not be supported by the bell, coupling, or any other joint surface. The pipe should rest on level ground and should not rest on rocks, boulders, or other hard debris that may cause a point load sufficient to gouge, crack, puncture, or otherwise damage the pipe wall. The pipe interior and all joining surfaces should be kept free of dirt and foreign matter.

Ultraviolet (UV) protection. Check with the pipe manufacturer regarding the necessity of UV protection when stored outside.

Nested pipe. Store nested pipe only in the original transport packaging. Do not stack nested pipe unless approved by the manufacturer. Transport pipe only in the original transport packaging.

10.5 REPAIR

Typically, damaged pipe can be repaired quickly and easily by qualified personnel at a job site. The repair design depends on the wall thickness, wall composition, application, and the type and extent of damage. Do not attempt to repair damaged or defective pipe without consulting the pipe manufacturer.

144 FIBERGLASS PIPE DESIGN

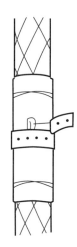

Figure 10-9 Patch

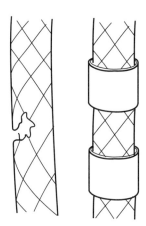

Figure 10-10 Cut out and replace

Source: Flowtite Technology, Sandefjord, Norway.

Figure 10-11 Steel coupling

Scrapes and gouges on the pipe exterior that are less than 10% of the pipe wall thickness generally require no repair, while deeper scrapes generally require repair. Repair for damage to the inner liner depends on the damage depth. Scratches, scrapes, and abrasion that do not penetrate through the entire liner generally require no repair. Gouges through the entire liner that penetrate the interior reinforced structural glass require a mat/resin lay-up to restore the original pipe wall thickness. Structural fracture of the pipe wall is evaluated on a case-by-case basis to provide repair sufficient to restore the original pipe strength.

Damaged pipe can either be replaced or repaired. During repair, the pipeline cannot be under pressure and the area to be repaired must be dry throughout the procedure. Repair techniques include patching small areas (Figure 10-9), cut out and replace (Figure 10-10), repair clamps, hand lay-up, and flexible steel couplings (Figure 10-11). Consult the pipe manufacturer to determine which method is appropriate.

Consult the pipe manufacturer regarding minor repairs of damaged fittings. Extensively damaged pipe and fittings must be replaced.

Hand lay-up repair. The pipe manufacturer should be contacted for job-specific lay-up instructions. Some manufacturers provide field lay-up kits individually prepared for the pipe diameter and pressure rating. Kits include premeasured resin, catalyst, and precut glass mat. The ambient temperature should be between 60°F and 100°F (15°C and 38°C), and the repair should be protected from the sun while curing to prevent temperature differentials. Lay-up repairs require clean, controlled conditions and skilled, trained personnel.

Flexible steel couplings can be used for joining pipe sections as well as for repairs. Steel repair couplings consist of a steel mantle with an interior rubber sleeve.

This page intentionally blank.

Glossary

Fiberglass pipe materials, processes, product standards, test methods, and installation practices and procedures may introduce some terms and terminology that are new to the unfamiliar user. This glossary provides basic definitions of fiberglass pipe terms used in this manual and by those in the fiberglass pipe industry.

accelerator See *hardener*.

adapter A fitting used to join two pieces of pipe, or two pipe fittings, that have different joining systems.

aggregate Siliceous sand conforming to ASTM C33, except that the requirements for gradation do not apply.

aliphatic amine curing agent Aliphatic amines are curing agents for epoxy resins. Aliphatic amine cured epoxy resins cure at room temperature, a property that makes them especially suitable for use in adhesives. Some filament-wound pipes use aliphatic amine cured epoxy resins. The properties of these pipes depend on the specific amine used in manufacture.

anhydride curing agents Anhydrides are widely used curing agents for fiberglass reinforced epoxy pipe. The properties of these pipes depend on the specific anhydride used in manufacture.

bell-and-spigot A joining system in which two cylindrical surfaces come together to form a seal by adhesive bonding or by compression of an elastomeric gasket. The bell is the female end; the spigot is the male end.

bisphenol-A A major ingredient used to make the most common type of epoxy resin, bisphenol-A epoxy resin. Also used as an intermediate to produce some polyester resins.

box The female end of a threaded pipe, or fitting, connection.

buckling See *collapse*.

burst pressure The ultimate pressure a pipe can resist for a short term before failing. Also see *weeping*.

burst strength (hoop stress) The circumferential stress (hoop stress) at burst pressure.

bushing A fitting used to join two different sizes of pipe by reducing the size of the female end of the joint.

catalyst See *hardener*.

centrifugal casting A process used to manufacture tubular goods by applying resin and reinforcement to the inside of a mold that is rotated and heated, subsequently polymerizing the resin system. The outside diameter of the finished pipe is determined by the inside diameter (ID) of the mold tube. The ID of the finished pipe is determined by the amount of material introduced into the mold. Other materials may be introduced in the process during manufacture of the pipe.

collapse Failure caused as the result of application of a uniform force around the total circumference of the pipe. The force may be caused by an externally applied pressure or vacuum inside the pipe. The mode of failure is usually stability related and occurs as flattening of the pipe but can be caused by compressive (shear) failure of the pipe wall.

collar See *coupling*.

compressive force The force that occurs when opposing loads act on a material, thus crushing or attempting to crush it. In pipe, circumferential compressive forces may result from external pressure; longitudinal compressive forces may result from heating of an end-restrained fiberglass pipe.

coupling (collar) A short, heavy-wall cylindrical fitting used to join two pieces of the same sized pipe in a straight line. The coupling always has female connection ends that can be threaded or that use adhesive bonding or elastomeric seals.

creep Deformation or strain that occurs over time when a material experiences sustained stress. Creep is expressed in inch per inch (millimeter per millimeter) per interval of time. Fiberglass pipe is subject to creep at all temperatures when subjected to stress.

cure The hardening of a thermoset resin system by heat and/or chemical action.

cure stages The degree to which a thermoset resin has cross-linked. In order of increasing cross-linking, the three stages include B-stage, gelled, and fully cured.

curing agent See *hardener*.

cut and mitered fittings Fittings manufactured by cutting, assembling, and bonding pipe sections into a desired configuration. The assembled product is then overlayed with resin-impregnated roving, mat, or glass cloth to provide required strength.

cyclic pressure rating The pressure rating obtained as the result of performing tests in accordance with ASTM D2992, procedure A. This method rates pipe on the basis of 150 million cycles. This conservative approach results in lower pressure ratings for pipes than static testing, but is useful in comparing competitive products.

design factor (factor of safety) A number equal to or greater than 1.0 that takes into consideration the variables and degree of safety involved in a design. Test data are divided by the design factor to obtain design allowable values. It is the reciprocal of the *service factor*. Also called *safety factor*.

drift diameter A measure of the effective minimum inside diameter of a pipe including ovality and longitudinal warpage over a given length of pipe.

elastic limit See *proportional limit*.

elastic modulus (modulus of elasticity) The "resistance" of a material to movement. The slope of the stress–strain curve within the elastic range.

epoxy resin (thermosetting) A polymer containing two or more three-membered rings, each consisting of one oxygen and two carbon atoms. The polymer is cured by cross-linking with an amine or anhydride hardener, with or without heat, catalyst, or both.

fatigue Permanent structural damage in a material subjected to fluctuating stress and strain.

fiberglass pipe A tubular product containing glass-fiber reinforcements embedded in or surrounded by cured thermosetting resin. The composite structure may contain aggregate, granular, or platelet fillers; thixotropic agents; and pigments or dyes. Thermoplastic or thermosetting liners or coatings may be included.

filament winding A process used to manufacture tubular goods by winding continuous glass-fiber roving or roving tape onto the outside of a mandrel or core pipe liner in a predetermined pattern under controlled tension. The roving may be saturated with liquid resin or preimpregnated with partially cured resin. Subsequent polymerization of the resin system may require application of heat. The inside diameter of the finished pipe is fixed by the mandrel diameter or the inner diameter of the core pipe liner. The outside diameter of the finished pipe is determined by the amount of material that is wound on the mandrel or core pipe liner. Other materials may be introduced in the process during the manufacture of the pipe.

fillers Extender materials added to a resin that do not affect the cure of the resin but may influence the physical and mechanical properties of the resin system and the finished product.

fitting types The classification of fittings by the method of manufacture (i.e., molded, cut and mitered, filament wound).

gel time The time it takes for a resin system to increase in viscosity so that flow will not occur.

glass fabric A bi-directional reinforcing material made by weaving glass-fiber yarn.

glass fibers A commercial grade of glass filaments with binder and sizing that are compatible with the impregnating resin.

hand lay-up Any of a number of manual methods for forming resin and fiberglass into finished pipe products. These procedures include overwrap techniques, contact molding, and hand molding. Complex shapes can be fabricated.

hardener (accelerator, catalyst, curing agent, promoter) Any of a number of chemicals added to the resin, individually or in combination, that speed up the curing process or cause hardening to occur.

hoop stress Circumferential stress. See also *burst strength*.

hydrostatic design basis (HDB) The long-term hydrostatic hoop strength of a specific fiberglass pipe material for water service as determined by tests and detailed evaluation procedures in accordance with ASTM D2992.

integral joint A joint configuration in which the connection is an integral part of the pipe. A length of pipe with integral joints will have one male end and one female end.

isopolyester Unsaturated polyester based on isophthalic acid.

joining (connecting) systems Any of a variety of methods for connecting two separate components of a piping system. Included are bell-and-spigot, threaded, coupling, and mechanical devices.

joint A term used to describe an individual length of pipe as well as the actual joining mechanism (i.e., adhesive-bonded bell-and-spigot, threaded, gasketed bell-and-spigot, gasketed coupling, etc.).

liner A filled or unfilled thermoplastic or thermosetting resin layer, nonreinforced or reinforced, forming the interior surface of the pipe.

matrix The resin material used to bind reinforcements and fillers together. This resin may be epoxy or polyester and, to a large extent, dictates the temperature and chemical performance for a pipe or fitting.

minimum bending radius The allowable deflection of the centerline of a pipe before damage occurs. The radius refers to an imaginary circle of which the pipe length would be an arc.

mitered fittings See *cut and mitered fittings*.

modulus of elasticity See *elastic modulus*.

molded fittings Pipe fittings formed by compressing resin, chopped fiber, and other ingredients in a mold under heat and pressure.

pin The male end of a pipe or fitting that matches with the female end of another pipe or fitting.

pipe stiffness A measure of the force required to deflect the diameter of a pipe ring a unit amount.

Poisson's effect (ratio) The property of a material that causes a change in its dimensions due to a force applied perpendicular to the plane of the dimension change. Expressed as the ratio of lateral strain to load direction strain.

polyester resin (thermosetting) An ethylenic unsaturated polymer with two or more ester groups, dissolved in a reactive diluent with vinyl unsaturation. The polymer is cured through cross-linking by means of a free-radical–initiated curing mechanism, such as peroxide catalyst and heat.

Any of a large family of resins that are normally cured by cross-linking with styrene. The physical and chemical properties of polyester resins vary greatly. Some have excellent chemical and physical properties, while others do not. Vinyl esters are a specific type of polyester resin. Polyester resins with properties suitable for use in the manufacture of fiberglass pipe include orthophthalic, isophthalic, bisphenol-A fumarate, and chlorendic anhydride acid polyesters. Each type of resin has particular strengths and weaknesses for a given piping application.

pressure class The maximum sustained pressure for which the pipe is designed.

pressure rating The maximum long-term operating pressure a manufacturer recommends for a given product. Also referred to as *design pressure*.

promoter See *hardener*.

proportional (elastic) limit The greatest stress a material can sustain for a short time without causing permanent deformation. It is defined by the point at which the stress–strain curve deviates from linearity. For composite materials, this point is called the apparent elastic limit since it is an arbitrary approximation on a nonlinear stress–strain curve. See *stress–strain diagram*.

reducer A pipe fitting used to join two different-sized pieces of pipe. With the same centerline in both pipes, the reducer is concentric; if centerlines are offset, it is eccentric.

reinforced polymer mortar pipe (RPMP) A fiberglass pipe with aggregate.

reinforced thermosetting resin pipe (RTRP) A fiberglass pipe without aggregate.

reinforcement Glass fibers used to provide strength and stiffness to a composite material. The form of reinforcement plays a major roll in determining the properties of a composite. The fiber diameter and the type of sizing used are also factors. Terms relating to the physical form of the reinforcement include:

>   *Chopped fiber*—Continuous fibers cut into short (0.125 in. to 2.0 in. [3.2 mm to 50 mm]) lengths.
>
>   *Filament*—A single fiber of glass (e.g., a monofilament).

Mats—A fibrous material consisting of random-oriented, chopped, or swirled filaments, loosely held together with a binder.

Milled fibers—Glass fibers, ground or milled into short (0.032 in. to 0.125 in. [0.81 mm to 3.2 mm]) lengths.

Roving—A collection of parallel glass strands or filaments coated with a finish or coupling agent to improve compatibility with resins, gathered without mechanical twist. Roving may be processed in a continuous or chopped form.

Yarn—Glass-fiber filaments twisted together to form textile type fibers.

Yield—The number of yards of material made from one pound of product.

resin Any class of solid or pseudosolid organic materials, often of high molecular weight, with no definite melting point. In the broad sense, the term is used to designate any polymer that is a basic material for plastics.

service factor A number less than or equal to 1.0 that takes into consideration the variables and degree of safety involved in a design. The service design factor is multiplied by test values to obtain design allowables. It is the reciprocal of the *design factor*.

static pressure rating The recommended constant pressure at which pipe can be operated continuously for long periods without failure. Determined by conducting tests in accordance with ASTM D2992, procedure B.

stiffness class The nominal stiffness of a specified pipe.

strain Dimensional change per unit of length resultant from applied force or load. Measured in inch per inch (millimeter per millimeter).

stress The force per unit of cross-sectional area. Measured in per square inch (kilopascals).

stress–strain diagram A graphic presentation of unit stress versus the corresponding unit strain. As the load increases, elongation or deformation of the material also increases.

support spacing The recommended maximum distance between pipe supports to prevent excessive pipe deformation (bending).

surface layer A filled or unfilled resin layer, nonreinforced or reinforced, applied to the exterior surface of the pipe structural wall.

surfacing mat A thin mat of fine fibers used primarily to produce a smooth surface on a reinforced plastic. Also called surfacing veil.

surge allowance That portion of the surge pressure that can be accommodated without changing pipe pressure class. The surge allowance is expected to accommodate pressure surges usually encountered in typical water distribution systems.

surge pressure A transient pressure increase greater than working pressure, sometimes called water hammer, that is anticipated in a system as a result of a change in the velocity of the water, such as when valves are operated or when pumps are started or stopped.

tape A unidirectional glass-fiber reinforcement consisting of rovings knitted or woven into ribbon form.

tensile force A force applied to a body tending to pull the material apart.

thermal conductivity The rate at which a material transmits heat from an area of high temperature to an area of lower temperature. Fiberglass pipe has low thermal conductivity.

thermal expansion The increase in dimensions of a material resulting from the application of heat. Thermal expansion is positive as temperature increases and negative as temperature decreases.

thermoplastic resin A plastic that can be repeatedly softened by heating and hardened by cooling and that, in the softened state, can be fused or shaped by flow.

thermoset A polymeric resin cured by heat or chemical additives. Once cured, a thermoset resin becomes essentially infusible (cannot be remelted) and insoluble. Thermosetting resins used in pipe generally incorporate reinforcements. Typical thermosets include:

> *epoxies*
> amine cured
> anhydride cured
> aliphatic polyanhydrides
> cycloaliphatic anhydrides
> aromatic anhydrides
>
> *novolac or epoxy novolac*
>
> *unsaturated polyesters*
> orthophthalic polyester
> isophthalic polyester
> bisphenol-A fumarate polyester
> chlorendic acid polyester
>
> *vinyl esters*
> bisphenol-A methacrylates
> bisphenol-F methacrylates

thrust forces Commonly used to describe the forces resulting from changes in direction of a moving column of fluid. Also used to describe the axial or longitudinal end loads at fittings, valves, etc., resulting from hydraulic pressure or thermal expansion.

torque Used to quantify a twisting force (torsion) in pipe. Torque is measured as a force times the distance from the force to the axis of rotation. Torque is expressed in foot-pounds (ft-lb) or inch-pounds (in.-lb) (Newton meters [N-m]).

ultimate pressure The ultimate pressure a pipe can resist for a short time before failing. This pressure is typically determined by the ASTM D1599 test. May also be referred to as ultimate burst pressure. When some fiberglass pipes are pressured to their ultimate pressure, the failure mode may be by leakage or weeping through the pipe wall rather than fracture of the pipe wall.

vinyl ester A premium resin system with excellent corrosion resistance.

weeping Leakage of minute amounts of fluid through the pipe wall.

working pressure The maximum anticipated, long-term operating pressure of the water system resulting from normal system operation.

woven roving A glass-fiber fabric reinforcing material made by the weaving of glass-fiber roving.

Index

NOTE: *f.* indicates figure; *t.* indicates table.

AASHTO LRFD Bridge Design Specification, 53
Aboveground pipe design and installation, 105
   allowable tensile or compressive loads, 118
   anchors, 111–112, 112*f.*
   bending, 116–117, 123
   and bending loads, 118
   design examples, 120–124
   and design pressure or stress, 118
   directional changes, 110–111, 111*f.*, 122
   expansion joints, 108, 108*f.*
   expansion loops, 109–110, 109*f.*, 122
   guides, 111, 111*f.*
   heat tracing, 117–118, 124
   and modulus of elasticity, 118
   and pipe properties and characteristics, 118–119
   and Poisson's ratio, 119
   spacing (anchors and guides), 107, 121
   supports, 112–115, 113*f.*, 115*f.*, 115*t.*, 116*f.*
   thermal conductivity, 117
   thermal end loads, 106–107, 120–121
   and thermal expansion and contraction, 105–111
   and vacuum or external pressure, 119
   See also Underground installation
Abrasion resistance, 13
Adhesive-bonded joints, 126, 127*f.*
   assembly, 134–135
   wrapped, 88
Aggregates
   manufactured, 78
   open-graded, 78
   processed, 78
   recommendations for use based on stiffness category and location in trench, 79, 81*t.*
   See also Backfill
Anchors, 107, 111–112, 112*f.*
ANSI/AWWA Standard C950, 3, 67, 75
Approvals, 5
Axial stress-strain curves, 15, 16*f.*

Backfill
   around angularly deflected pipe joints, 89–90
   cementitious backfill materials, 83
   compaction of soils with few fines (SC1, SC2), 88–89
   compaction of soils with significant fines (SC3, SC4, SC5), 89
   compaction of soils with some fines (SC2), 89
   compaction under haunches, 88, 89*f.*
   compatibility of pipe and, 82
   densification using water, 89
   determination of in-place density of soils, 89
   maximum particle size, 82
   migration, 82–83
   minimum cover, 90
   minimum density, 89
   moisture content, 80–82
   See also Aggregates, Soil
Bedding, 77
   coefficient, 47, 53
Bell-and-spigot joints, 125, 127*f.*, 126–128, 129*f.*, 130*f.*, 131*f.*
Bending loads, 118
Bends and bending, 93, 94*f.*, 116–117, 123
   avoiding excessive, 115
   design factor, 51
   ring bending, 50–51
Bifurcations, 94*f.*
Biological-attack resistance, 13
Bonded joints. *See* Adhesive-bonded joints
Buckling
   calculations, 65–66
   scaling constant, 119, 122
   theory, 65
Buried pipe design, 43
   and axial loads, 67
   and buckling, 65–66
   calculations and requirements, 47–66
   and combined loading, 60–65
   conditions, 46
   definitions, 43–44
   design factor, 44
   example, 67–74, 68*t.*
   and head loss, 46
   and installation parameters, 47
   and internal pressure, 47–50
   and pipe properties, 46–47
   procedure, 47
   and ring bending, 50–51
   special considerations, 67
   and surge pressure, 46, 49–50
   symbols, 44–45
   See also Aboveground pipe design and installation, Underground installation

Centrifugal casting, 22–23
   chopped glass reinforcement method, 23*f.*, 24
   preformed glass reinforcement sleeve method, 22*f.*, 23–24
Chemical resistance, 12
Circumferential stress-strain curves, 15, 15*f.*

Colebrook equation, 30
Combined loading, 60–65
Compactibility, 77
Compression molding, 135–136, 136f.
Compressive scaling constant, 119
Constrained soil modulus, 59–60, 62t.–63t.
   for the native soil, 47, 59–60, 64t.
Contact molding, 137
Corrosion resistance, 9
Cut and miter process, 136f., 137, 137f., 138f.
Cyclic pressure testing, 16f., 17

Darcy-Weisbach equation, 29–30
Dead ends, 93, 94f.
Deflection, 77
Deflection lag factor, 47, 53
Design
   and deflection, 51–60
   and pipe properties, 46–47
   See Aboveground pipe design and
      installation, Buried pipe design
Design factor, 44
Design stress, 118
D_f. *See* Shape factor
Dimensional stability, 9
Directional changes, 110–111, 111f., 122
D_L. *See* Deflection lag factor

Electrical properties, 9
Elevation head, 36
Embedment materials. *See* Aggregates,
      Backfill, Pipe zone embedment, Soil
Engineer, defined, 77
Expansion joints, 108, 108f.
Expansion loops, 109–110, 109f., 122
External pressure, 119

Fiberglass
   accelerators, 12
   composition, 1, 9, 10–12
   fillers, 12
   glass fiber reinforcements, 10–11
   inhibitors, 12
   pigments, 12
   promoters, 12
   resins, 11–12
Fiberglass couplings, 125
Fiberglass pipe, 1
   abrasion resistance, 13
   applications, 2
   axial stress-strain curves, 15, 16f.
   characteristics, 9–10
   chemical resistance, 12
   circumferential stress-strain curves, 15, 15f.
   composition, 1, 9, 10–12
   flame retardants, 13
   history, 1–2
   low fluid resistance, 25
   mechanical property range, 14t., 15
   mechanical property testing, 15–17, 15f., 16f.
   non-tuberculation, 15
   note on terminology, 7
   physical properties, 12–15
   properties and characteristics, 9–10,
      12–15, 46–47, 118–119
   resistance to biological attack, 13
   standards, 3–7
   static vs. cyclic pressure testing, 15–17, 16f.
   temperature resistance, 12–13
   types, 1
   weathering resistance, 13
Filament winding, 19, 20f., 136–137
   continuous advancing mandrel method, 21,
      21f., 22f.
   continuous methods, 21
   finished pipe, 21, 22f.
   mandrels, 19, 20f.
   multiple mandrel method, 21
   reciprocal method, 19
   ring and oscillating mandrel method, 21
Final backfill, 77
Fines, 77
Fittings and specials, 135
   compression molding, 135–136, 136f.
   contact molding, 137
   cut and miter process, 136f., 137, 137f., 138f.
   filament winding, 136–137
Flame retardants, 13
Flanges, 130–132, 132f., 133f.
   assembly, 135
Flow rate, conversion to fluid velocity, 26
Fluid velocity equation, 26
   examples, 36, 39
Foundation, 77
Friction factor. *See* Moody friction factor
FS. *See* Design factor

Gasketed coupling joints, 128, 130f., 131f.
Gaskets, 126
Gasket-sealed joints, 126–130, 129f., 130f.
Geotextile, defined, 77
Glass fiber reinforcements, 10
   arrangements, 11
   bidirectional, 11
   continuous roving, 10
   forms, 10–11
   multidirectional (isotropic), 11
   reinforcing mats, 10
   surface veils, 11
   types, 10
   unidirectional, 11
   woven roving, 10
Guides, 107, 111, 111f.

Handling, 139, 140
   bundles (unitized loads), 140, 141f., 142f.
   nested pipe, 141–142, 142f., 143f.
   with single and double slings, 140, 141f.

Haunching, 77
Hazen-Williams equation, 27–28, 27f.
    to calculate head loss in fittings, 30–32
    example, 35
    simplified, 28
HDB. *See* Hydrostatic design basis
Head loss, 27, 46
    calculation examples, 36, 37
    conversion to pressure loss, 28, 36, 40
    conversion to pump horsepower demand, 33, 37
    in fittings, 30–32
Heat tracing, 117–118, 124
Hydrodynamic thrust, 93
Hydrostatic design basis, 44
Hydrostatic thrust, 93–94, 94f.

In situ soils, 78, 79
Initial backfill, 77
Installation
    parameters, 47
    split, 78
    See also Aboveground pipe design and installation, Underground installation

Joints, 88, 125
    adhesive-bonded, 126, 127f.
    adhesive-bonded and wrapped, 88
    angularly deflected, 88
    assembly, 132–135
    backfill around angularly deflected joints, 89–90
    bell-and-spigot, 125, 127f., 126–128, 129f., 130f., 131f.
    elastomeric seal (gasketed), 88
    expansion, 108, 108f.
    fiberglass couplings, 125
    flanged (assembly), 135
    gasketed coupling, 128, 130f., 131f.
    gaskets, 126
    gasket-sealed, 126–130, 129f., 130f.
    mechanical (flanges), 130–132, 132f., 133f.
    mechanical coupling, 125, 133f.
    reinforced-overlay, 126, 128f., 129f.
    restrained (tied), 99–103, 101f., 103f., 125–126
    restrained-gasketed, 128–130, 131f.
    safety precautions in assembly, 135
    with small horizontal deflections, 97–99, 98f., 100f.
    with small vertical deflections, 99, 101f.
    threaded (assembly), 135
    tied, at horizontal bends and bulkheads, 99–102, 101f.
    tied, at vertical (uplift) bends, 102–103, 103f.
    unrestrained, 125

K factors, 30–32, 32t.
    example, 40

K_x. *See* Bedding coefficient

Laterals, 93
Lightness (weight), 9
Live loads on the pipe, 53–56, 54f., 57t., 58f., 59f.
Long-term, ring-bending strain, 51
Loss coefficients (K factors), 30–32, 32t.
    example, 40

Maintenance cost, 9
Manning equation, 28–29
    to calculate head loss in fittings, 30–32
Manufactured aggregates, 78
Manufacturing processes. *See* Centrifugal casting, Filament winding
Maximum standard Proctor density, 78
Maximum velocity (equation), 25
Maximum velocity for corrosive or erosive fluids (equation), 26
Mechanical coupling joints, 125, 133f.
Mechanical joints, 130–132, 132f., 133f.
Minimum pipe diameter for corrosive or erosive fluids (equation), 26
Minimum pipe diameter for water (equation), 26
    example, 35–36
Mitered fittings, 136f., 137, 137f., 138f.
Modulus of elasticity, 118
Moody friction factor, 30
    diagram, 30, 31f.
    examples, 36, 40
M_p. *See* Multiple presence factor
M_s. *See* Constrained soil modulus
M_{sb}. *See* Pipe backfill surround, Pipe zone embedment
M_{sn}. *See* Constrained soil modulus: for the native soil
Multiple presence factor, 53–54

Native (in situ) soil, 78
Nonconductivity, 9

Open-graded aggregate, 78
Optimum moisture content, 78

P_c. *See* Pressure class
Pipe backfill surround, 59
Pipe diameters
    calculating, 26
    calculation example, 35–37
Pipe installation. *See* Installation
Pipe properties, 46–47
Pipe sizing equations, 25–26
Pipe stiffness, 56–57
Pipe zone embedment, 47, 78
Pipeline energy consumption calculation, 32–33
    example, 37–38

Piping codes, 5–6
Poisson's ratio, 119
Pressure class, 44, 47–49
   calculation example, 35–37
Pressure pipe, typical diameters (equation), 26
Pressure reduction calculations, 27–30
Pressure surge, 34
   calculating (Talbot equation), 34–35
   calculation example, 38–41
   cause and control of, 34
Pressure testing, 15–17, 16f., 91–92
Processed aggregates, 78
Product listings, 5
Properties and characteristics, 9–10, 12–15
   electrical, 9
   related to aboveground pipe design, 118–119
   related to buried pipe design, 46–47
PS. *See* Pipe stiffness
P_s. *See* Surge pressure
P_{sa}. *See* Surge allowance
P_w. *See* Working pressure

Reducers, 93, 94f.
Reinforced-overlay joints, 126, 128f., 129f.
Relative density, 78
Repair, 139, 143–145
   clamps, 145
   cut out and replace, 144f., 145
   hand lay-up, 145
   patching, 144f., 145
   steel couplings, 144f., 145
Resins, 10, 11
   epoxy, 11–12
   polyester, 11
Restrained joints, 125–126
Restrained-gasketed joints, 128–130, 131f.
Reynolds number, 29–30
   equation, 29
   examples, 36, 40
Ring bending, 50–51

S_b. *See* Long-term, ring-bending strain
S_c. *See* Soil support combining factor
Scaling constants, 119
Service line connections, 137
Shape factor, 50–51, 51t.
Shipping, 139–140, 140f.
Soil
   classification chart, 61t.
   compaction recommendations by stiffness category, 88–89
   constrained soil moduli for the native soil, 47, 59–60, 64t.
   constrained soil modulus, 59–60, 62t.–63t.
   determination of in-place density, 89
   horizontal bearing strengths, 96, 96t.
   in situ, 78, 79
   recommendations for use based on stiffness category and location in trench, 79, 81t.
   stiffness, 78
   stiffness classes, 79–80, 80t.
   support combining factor, 59–60, 64t.
   vertical soil load on the pipe, 53
   See also Backfill
Specials. *See* Fittings and specials
Split installation, 78
Standards
   ISO, 6–7
   organizations issuing, 2
   product specifications and classifications, 3–4
   recommended practices, 4
   standard test methods, 4–5
Static pressure testing, 16f., 17
Storage, 139, 142–143
   of nested pipe, 143
   stacking, 142, 143f.
   ultraviolet protection, 143
Strength-to-weight ratio, 9
Stress-strain curves, 15, 15f., 16f.
Suction pipe, typical diameters (equation), 26
Supports, 112, 113f.
   avoiding excessive bending, 115
   avoiding point loads, 114
   and heavy equipment, 115, 116f.
   minimum support widths, 114, 115t.
   protecting against external abrasion, 114, 115f.
   Type I, 112–113
   Type II, 113
   Type III, 114
   Type IV, 114
   vertical, 115, 116f.
Surge allowance, 44
Surge pressure, 44, 46, 49–50
Symbols, 44–45

Talbot equation, 34–35
   examples, 39, 40
Tees, 93, 94f.
Temperature resistance, 12–13
Tensile or compressive loads, 118
Thermal conductivity, 117
Thermal end loads, 106–107, 120–121
Thermal expansion and contraction, 105–106
Thrust, 93–94
Thrust blocks, 95
   and adjacent excavation, 97
   and horizontal bearing strengths of soils, 96, 96t.
   for horizontal bends, 95, 95f., 96
   proper construction, 96–97
   size calculation, 95–97, 96t.
   for vertical bends, 96, 96f.

Thrust resistance, 94
  joints with small horizontal deflections, 97–99, 98f., 100f.
  joints with small vertical deflections, 99, 101f.
  restrained (tied) joints, 99–103, 101f., 103f.
  thrust blocks, 95–97, 95f., 96t., 97f.
  tied joints at horizontal bends and bulkheads, 99–102, 101f.
  tied joints at vertical (uplift) bends, 102–103, 103f.
  transmission of thrust force through pipe, 103
Trenches and trenching
  bedding material, 86
  bedding support, 86, 86f.
  bottom level and stability, 85, 86
  excavation, 83–84
  foundation, 86
  localized loadings, 86, 87f.
  minimum trench width, 84
  overexcavation, 88
  and rock or unyielding material, 86
  on slopes, 85
  sloughing, 88
  wall supports, 84
  and water control, 83
Tuberculation, 15

Underground installation, 75, 85–86
  adhesive bonded and wrapped joints, 88
  angularly deflected joints, 88
  ASTM standards related to, 76–77
  contract documents, 92
  deflection monitoring, 91
  and differential settlement (manholes, rigid structures, changing foundation soils), 90
  elastomeric seal (gasketed) joints, 88
  exposing pipe for service line connections, 91
  field monitoring, 91–92
  jointing, 88
  location and alignment, 88
  parallel piping systems, 91
  pipe caps and plugs, 91
  placing and compacting backfill, 88–90
  placing and joining pipe, 88
  and pressure testing, 91–92
  terminology, 77–78, 78f.
  trench preparation, 86–88
  and vertical risers, 90
  See also Aboveground pipe design and installation, Aggregates, Backfill, Soil, Thrust blocks
Unrestrained joints, 125

Vertical soil load on the pipe, 53

Water hammer. *See* Pressure surge
W_c. *See* Vertical soil load on the pipe
Weathering resistance, 13
W_L. *See* Live loads on the pipe
Working pressure, 43, 49–50
  calculation example, 35–37, 40
Wyes, 94f.

This page intentionally blank.

AWWA Manuals

M1, *Principles of Water Rates, Fees, and Charges,* Fifth Edition, 2000, #30001PA

M2, Instrumentation and Control, Third Edition, 2001, #30002PA

M3, *Safety Practices for Water Utilities,* Sixth Edition, 2002, #30003PA

M4, *Water Fluoridation Principles and Practices,* Fifth Edition, 2004, #30004PA

M5, *Water Utility Management Practices,* First Edition, 1980, #30005PA

M6, *Water Meters—Selection, Installation, Testing, and Maintenance,* Fourth Edition, 1999, #30006PA

M7, *Problem Organisms in Water: Identification and Treatment,* Third Edition, 2004, #30007PA

M9, *Concrete Pressure Pipe,* Second Edition, 1995, #30009PA

M11, *Steel Pipe—A Guide for Design and Installation,* Fifth Edition, 2004, #30011PA

M12, *Simplified Procedures for Water Examination,* Third Edition, 2002, #30012PA

M14, *Recommended Practice for Backflow Prevention and Cross-Connection Control,* Third Edition, 2003, #30014PA

M17, *Installation, Field Testing, and Maintenance of Fire Hydrants,* Third Edition, 1989, #30017PA

M19, *Emergency Planning for Water Utility Management,* Fouth Edition, 2001, #30019PA

M21, *Groundwater,* Third Edition, 2003, #30021PA

M22, *Sizing Water Service Lines and Meters,* Second Edition, 2004, #30022PA

M23, *PVC Pipe—Design and Installation,* Second Edition, 2002, #30023PA

M24, *Dual Water Systems,* Second Edition, 1994, #30024PA

M25, *Flexible-Membrane Covers and Linings for Potable-Water Reservoirs,* Third Edition, 2000, #30025PA

M27, *External Corrosion Introduction to Chemistry and Control,* Second Edition, 2004, #30027PA

M28, *Rehabilitation of Water Mains,* Second Edition, 2001, #30028PA

M29, *Water Utility Capital Financing,* Second Edition, 1998, #30029PA

M30, *Precoat Filtration,* Second Edition, 1995, #30030PA

M31, *Distribution System Requirements for Fire Protection,* Third Edition, 1998, #30031PA

M32, *Distribution Network Analysis for Water Utilities,* Second Edition, 2005, #30032PA

M33, *Flowmeters in Water Supply,* First Edition, 1989, #30033PA

M36, *Water Audits and Leak Detection,* Second Edition, 1999, #30036PA

M37, *Operational Control of Coagulation and Filtration Processes,* Second Edition, 2000, #30037PA

M38, *Electrodialysis and Electrodialysis Reversal,* First Edition, 1995, #30038PA

M41, *Ductile-Iron Pipe and Fittings,* Second Edition, 2003, #30041PA

M42, *Steel Water-Storage Tanks,* First Edition, 1998, #30042PA

M44, *Distribution Valves: Selection, Installation, Field Testing, and Maintenance,* First Edition, 1996, #30044PA

M45, *Fiberglass Pipe Design,* Second Edition, 2005, #30045PA

M46, *Reverse Osmosis and Nanofiltration,* First Edition, 1999, #30046PA

M47, *Construction Contract Administration,* First Edition, 1996, #30047PA

M48, *Waterborne Pathogens,* First Edition, 1999, #30048PA

M49, *Butterfly Valves: Torque, Head Loss, and Cavitation Analysis* First Edition, 2001, #30049PA

M50, *Water Resources Planning,* First Edition, 2001, #30050PA

M51, *Air-release, Air/Vacuum and Combination Air Valves,* First Edition, 2001, #30051PA

M54, *Developing Rates for Small Systems,* First Edition, 2004, #30054PA

To order any of these manuals or other AWWA publications, call the Bookstore toll-free at 1-800-926-7337.

This page intentionally blank.